Raghu Krishnaswamy.

The World
According to Wavelets

And what is the use of a book, thought Alice,
without pictures or conversations?

—Lewis Carroll, *Alice's Adventures in Wonderland*

The World According to Wavelets

The Story of a Mathematical Technique in the Making

Second Edition

Barbara Burke Hubbard

A K Peters
Wellesley, Massachusetts

Editorial, Sales, and Customer Service Office

A K Peters, Ltd.
289 Linden Street
Wellesley, MA 02181

Library of Congress Cataloging-in-Publication Data

Hubbard, Barbara Burke,
      The world according to wavelets : the story of a mathematical technique in the making / Barbara Burke Hubbard. -- 2nd ed.
              p.   cm.
      Includes bibliographical references (p.   —   ) and index.
      ISBN: 1-56881-072-5
      1. Wavelets (Mathematics) I. Title.
QA403.3.H83   1998
515'.2433--dc21

                               97-48738
                                   CIP

Printed in the United States of America
02 01 00 99 98   10 9 8 7 6 5 4 3 2 1

In memory of my father,
Vincent J. Burke,
who taught me that one must love one's work

And for my mother,
Velma Whitgrove Burke,
who sees wonders in the commonplace

Contents

To the Reader

When at the age of four or five I asked my mother just how babies are made, her answer was so obviously absurd that I didn't believe her, although I knew she never lied to me. I had at times the same feeling while writing this book: statements that the experts seemed to find natural, even commonplace, seemed barely credible. What I have tried to do in this book is to make them simultaneously surprising and believable.

The project started when the executive editor of the National Academy Press asked me to write an article about wavelets for a book, *A Positron Named Priscilla*, on the "frontiers of science." At the time I knew nothing about Fourier analysis and had never heard of wavelets; my only mathematical qualification—non-negligible—was my husband, a mathematician at Cornell University. (I never took calculus in high school because I spent my senior year in Moscow, where my father was sent as a newspaper correspondent; in college I carefully avoided all math courses.)

Rashly, I agreed to the assignment, and embarked on a period of work both passionate and disconcerting, which has led me to think about the possibilities of communicating mathematical ideas. Mathematicians claim that math is not a spectator sport: you cannot understand math, or enjoy it, without doing it. A mathematician who tries to communicate his subject to the layman soon finds himself in trouble. "This is getting vaguer and vaguer," lamented Robert Strichartz of Cornell as he tried to explain function spaces to me. "When I stick to the truth I don't communicate, and when I communicate I stray from the truth."

Talking without being understood is pointless, lying is painful, so most often mathematicians abandon the attempt, if they have the courage to try in the first place. It would be a shame to leave it at that. There exists, certainly, a limit to what someone without the right background can understand of mathematics, but I am convinced that we are far from having reached that limit. Mathematics contains ideas that can be, and deserve to be, communicated to a wider public—even if what is communicated is at the level of appreciation rather than practical knowledge.

No one claims that only geneticists doing recombinant DNA research should know what DNA is, or that only physicists and chemists should know that matter is made of atoms. In mathematics the reverse is often the case; too often we insist that children and students learn practical techniques without showing them that these techniques are based on interesting ideas, or that they make it possible to tackle interesting questions. No wonder some parents and educators are appalled when a child uses a calculator to solve problems; to them, mathematics is synonymous with computations, and if this work is given to a machine, nothing is left.

Fourier analysis and wavelets are good subjects with which to try another approach. The idea that information can be represented in different forms—that a mathematical function or a physical signal can become a *Fourier transform* or a *wavelet transform*, and that this transformation is useful—has had an enormous impact, both intellectual and practical, on our society.

This book is then both modest in scope and daring (not to say foolhardy) in design. It is only a brief introduction to a vast subject that itself is only a small part of mathematics; but I wanted to make this subject accessible to those who have no mathematical background to speak of, while giving enough details so that those who are more sophisticated will find it useful.

I have written the book on two levels. The main article contains no formulas. When mathematicians, physicists, or electrical engineers talk about Fourier analysis or wavelets, sooner or later they scribble down formulas. To them, formulas are the most precise and economical way of expressing their thoughts; naturally they think they are communicating something. But showing a formula to someone who can't remember whether $\sum$ means sum or integral is like asking someone who doesn't know which note on the musical staff is B-flat and which is D to analyze the score for a four-voice fugue by Bach.

This reaction isn't limited to English majors like me. "Is there any book that deals with fractal geometry in plain English instead of integrals, sigma notation, and a whole bunch of other funny symbols that I haven't learned yet?" a computer science major who had studied calculus asked my husband.

And in 1857 Michael Faraday, best known for his pioneering work connecting magnetism and electricity, wrote 26-year-old James Clerk Maxwell, 40 years his junior,

When a mathematician engaged in investigating physical actions and results has arrived at his conclusions, may they not be expressed in common language as fully, clearly, and definitely as in mathematical formulae? If so, would it not be a great boon to such as I to express them so?—translating them out of their hieroglyphics, that we also might work upon them by experiment. I think it must be so, because I have always found that you could convey to me a perfectly clear idea of your conclusions, which, though they may give me no full understanding of the steps of your process, give me the results neither above nor below the truth, and so clear in character that I can think and work from them. If this be possible, would it not be a good thing if mathematicians, working on these subjects, were to give us the results in this popular, useful, working state, as well as in that which is their own and proper to them [Faraday, p. 206]?

In the beginning, formulas can seem more of a hindrance than a help. If they are to serve any purpose they must be translated into words, but the mere sight of a formula on the page can be paralyzing. So I have not used "funny symbols" or "hieroglyphics" in the main article. (The expression $f(x) = x^2$ slipped in, but you can ignore it.)

On the other hand, formulas were not invented simply as weapons of intimidation. Although it may seem paradoxical, it is easier to understand Fourier analysis and wavelets if you see how it actually works, rather than clinging to generalities and metaphors. (More interesting, too—as in the case of making babies...) But to explain these details, words are often clumsy and sometimes ambiguous.

"When I use a word," says Humpty Dumpty in *Through the Looking Glass*, "it means just what I choose it to mean—neither more nor less." Following this creed, wavelet researchers sometimes attach new meanings to established words. *Dilate* can mean contract, *large-scale* can mean small-scale, *decimate* can mean cut in half. Even with standard definitions, sentences can be interpreted in unintended ways. Along with the Fourier transform and the wavelet transform, there exists a transform that changes spoken or written information into information understood by the listener or reader; with this transform—as I, to my chagrin, had many occasions to note—perfect reconstruction of the original signal is all too rare.

I have, then, included formulas in the articles you will find at the end of the book in the section *Beyond Plain English*, although they introduce two dangers. The first is that formulas have the nasty trick of exposing errors that might remain decently veiled if one would only remain sufficiently vague. "If you force me to be precise, of course I'll make mistakes," protested the well-known French mathematician René Thom in a lecture when his ungrateful audience first insisted on more precision, and then found mistakes in his formulas.

The other danger is that some readers may be frightened away. "When I see a formula, I start to panic; I always wonder, am I going to understand?" a colleague told me. There's no reason to be afraid; at the risk of irritating those who don't need it, everything is explained in detail. (Some remnants of high-school algebra and trigonometry are assumed, but if those remnants are frayed, some of the basics are reviewed in the Appendix. Calculus, while useful, is not essential.)

The Appendix is schizophrenic. It began as a place to hide things that might seem too daunting or that require mathematical techniques beyond the scope of this book: proofs of the Heisenberg uncertainty principle and the sampling theorem, for example. But readers who are less sophisticated mathematically will also find, in addition to a brief review of trigonometry, a list of mathematical symbols and a discussion of integrals.

The bibliography includes technical books and articles that may be useful to some readers. I have also included a list of wavelet software. Neither list claims to be complete. I have given detailed references where possible, but for many quotations no references are given; they come either from conferences organized by the National Academy of Sciences in Irvine, California, in November 1992, or (especially) from subsequent interviews and personal communications. When possible I have used published translations of French texts, but generally the translations are my own.

. . .

Too often, I think, mathematics is taught as a collection of final results and techniques divorced from human thought or feeling. This is misleading. Mathematics is rewarding, but fickle and unforgiving; in an instant one can have an inspired idea that illuminates an entire field, but one can also watch helplessly as an edifice of results arduously constructed over months or years crumbles because of one small hole that refuses to be patched, or, that when patched, reopens elsewhere. The would-be mathematician who does not love mathematics will not withstand the moments of doubt and the empty weeks or months when ideas don't come, or come only to prove false. And despite the public image of the solitary mathematician immersed in abstractions, mathematicians work as intimately with other people as musicians in a string quartet. Collaboration is the rule, not the exception, and that collaboration often crosses national boundaries.

But the human side of mathematics is generally well hidden from outsiders; mathematical writing, as Philip J. Davis and Reuben Hersh [Davis, Hersh] write in *The Mathematical Experience*, generally "follows an unbreakable convention: to conceal any sign that the author or the intended reader is a human being."

In teaching, too, generally one starts at the end: everything is settled and there is no hint of the often tortuous path that led to those results. The newcomer may feel stupid not to understand immediately, although mathematicians often spent years arriving at results and feeling at ease with them. The Greeks were profoundly troubled by irrational numbers; in the sixteenth century, negative numbers were considered "impossible" solutions; in the eighteenth century, long division was taught in universities. Even those not easily intimidated may not realize that mathematics is an ongoing process, that there are questions that have not yet been answered and questions that have not yet been asked. And they may not see what is beautiful and surprising in facts presented, with reticence, like so many historical dates to memorize.

To me the process is as interesting as the result; I see Fourier analysis and wavelets not only as tools but also as a story to tell, of ideas and people. I am deeply grateful to the researchers, in France and in the United States, who shared with me not just the results of their work but also its human context.

Nevertheless this approach is inaccurate. Even when talking only about mathematics, attributing discoveries is difficult. How can one evaluate the influence of a conversation or an article, of an answer—or a question—at the right moment, without even considering all the existing mathematics that any new result builds upon? "The names of theorems are distributed like the names of streets," the French mathematician André Weil is reputed to have said. Doing justice to all those who have contributed to the development of wavelets is yet more difficult, since wavelets were developed independently in different fields. I could not, in this limited framework, mention all the researchers who are part of this story, and who deserve to find their names here. I apologize to them.

Acknowledgments

I could never have written this book without help from a great many people. Two mathematicians in particular made crucial contributions, one who is among the foremost wavelet researchers, another who works in an entirely different field. Yves Meyer spent countless hours, first in a series of interviews, or rather private lessons, and then reading the text, making many corrections and valuable suggestions. I will be forever indebted to him for his help, enthusiasm, and encouragement. The moral, material, and intellectual support of my husband, John Hubbard, was also essential. In addition to serving as translator and guide to mathematical concepts and formulas and providing technical assistance with illustrations and TeX, he kept me honest when I lapsed into fuzzy thinking and vague language. Although he claims not to believe me, his mathematical explanations gave me a real pleasure, for which I thank him too.

The wavelets researchers whom I asked for help responded with extraordinary generosity. Honoring Fourier's tradition, they displayed "an inexhaustible patience," whether explaining their subject in an interview or answering a question that appeared out of the blue on their computer screen. I have come to admire not just their professional competence, but also their ingenuity in finding simple words to explain technical matters, and the courtesy with which they spared my self-esteem. The pleasure I got in writing this book is largely due to them, and I cannot thank them enough.

I owe special thanks to Ingrid Daubechies for the care with which she read an earlier version of the text and for her corrections and suggestions, as well as for her lucid and prompt answers to many questions; Stéphane Mallat, who cleared up many troublesome points, helped me to avoid numerous pitfalls, and in particular insisted on a less magical portrayal of multiresolution theory (never complaining when, as a result, I deluged him with ill-formulated questions); Olivier Rioul, who showed me that I had misunderstood certain things and helped me to understand them with such good will that I gladly pardoned him the frustration his honesty originally caused me; and Victor Wickerhauser for his endless good will and good humor. While answering countless questions, they helped to keep up my spirits during long months when I was all too aware that I was out of my depth.

It's also a pleasure to thank David Donoho for his patient answers and useful suggestions; Marie Farge, who when I first arrived in her office, thoroughly confused, put her work aside and spent two hours explaining things from the beginning; David Field for a glimpse into how wavelets were developed and are used outside mathematics and signal processing; Michael Frazier for his valiant if doomed attempt at Irvine to teach me everything I ever wanted to know about harmonic analysis in two hours, as well as for verifying certain formulas; Leonard Gross for his suggestions and corrections concerning quantum mechanics; Alex Grossmann, whose gift for metaphor enlivened a very instructive interview; Jean Morlet for a very interesting historical perspective; and Robert Strichartz for his explanations of function spaces and his advice during final revisions.

Others provided assistance by mail or e-mail for which I would like to thank them: Edward Adelson, Christophe d'Alessandro, Michel Barlaud, Jonathan Berger, Gregory Beylkin, Ronald Coifman, Karen De Valois, Russell De Valois, Ronald DeVore, Uriel Frisch, Jeffrey Geronimo, Eric Goirand, Dennis Healy, Shubha Kadambe, Richard Kronland-Martinet, Bradley Lucier, Gilbert Strang, Gene Switkes, Bruno Torrésani, Michael Unser, and Martin Vetterli.

The list, as Virginia Woolf wrote in the preface to *Orlando*, "... threatens to grow too long and is already far too distinguished. For while it rouses in me memories of the pleasantest kind it will inevitably wake expectations on the reader which the book itself can only disappoint." So I will conclude by thanking Stephen Mautner of the National Academy

Press, who first asked me to write about wavelets; Philippe Boulanger, who suggested expanding the original English article into a book in French and, when I told him I couldn't, insisted; Ralph Oberste-Vorth, Ricardo Oliva, and Dierk Schleicher for technical assistance; Yuval Fisher for help with illustrations; my brother, Douglas Burke, for pointing out sections that were unclear; Fred Kochman for helpful comments; T.W. Körner for his book *Fourier Analysis*, which contains stories that are both instructive and entertaining; the research staff at Olin Library, Cornell University; and SIAM for permission to reprint a figure. I would also like to thank my children, Alex, Eleanor, Judith, and Diana Hubbard, for helping to avert complete disaster to house and garden during the past few months, and Judith and Diana for waiting so patiently for promised dance and horseback riding lessons.

Much of the work was done in France while my husband worked at the Institut des Hautes Etudes Scientifiques in Bures-sur-Yvette; I would like to thank the institute for its hospitality, and especially Françoise Schmit, who helped me find historical documents and research papers. Last but not least I wish to thank all those at A K Peters for what I consider one of the most remarkable transformations, turning a manuscript into a book: in particular, Alice and Klaus Peters, Alexandra Benis, Iris Kramer-Alcorn, and Erin Miles.

A number of those who helped with the first edition did so again with this second edition. In particular, I would like to thank Stéphane Mallat for his explanations and for giving me the opportunity to read his own wavelet book before it was published, Ted Adelson, Jonathan Berger, Gregory Beylkin, Michael Frazier, Leonard Gross, Michael Unser, and Victor Wickerhauser.

In addition, the following people contributed to the second edition through their suggestions, corrections, or responses to inquiries: Felix Abramovich, Jean-Pierre Antoine, Norman Bleistein, Christopher Brislawn, Sidney Burrus, Roberto Marcondes Cesar Jr., Richard Clark, Catherine Cooper, Luciano da F. Costa, Jean-Philippe Couderc, J. J. Duistermaat, Piotr Durka, Kevin Englehart, Gabriel Fernández, Fernando Gomez, Amara Graps, Peter Heller, Felix Herrmann, Xiaoming Huo, Harm Jonker, Gerald Kaiser, Frank Kurth, Tom Hopper, Irving Kaplansky, Tom Lane,

Lars Lippert, François Meyer (who very kindly provided the brushlet illustration, as well as a clear explanation of his work), Colm Mulcahy, Ole M. Nielsen, Laurent Nottale, George Papanicolaou, Mary E. Patterson, Kannan Ramchandran, Vladimir Rokhlin, Mary Beth Ruskai, Matthias Schwab, Sergio Servetto, Jerry Shapiro, Eero Simoncelli, Wim Sweldens, Carl Taswell, Geert Uytterhoeven, and Hans van den Berg. I thank them for their help and apologize to anyone whose name was inadvertently omitted.

It should not be necessary to add that none of these people bears any responsibility for mistakes or omissions. Any virtues of this book have many authors, but its shortcomings are mine alone—especially as I did not always follow the good advice I was given. Since I wanted to write a book that would be accessible to readers with minimal mathematical background, I have chosen at times to follow my own intuition, reasoning that I am, regrettably, well placed to understand the needs of the mathematically innocent. The distinguished researchers who helped me may find it embarrassing to be associated with a text that is sometimes very elementary; I apologize to them for any discomfort and assure the reader that the unusual range of mathematical sophistication in this book is entirely my responsibility.

PART I
The World According to Wavelets

Mathematical Analysis

... mathematical analysis is as extensive as nature itself; it defines all perceptible relations, measures time, spaces, forces, temperatures; this difficult science is formed slowly, but it preserves every principle which it has once acquired....

Its chief attribute is clearness; it has no marks to express confused notions. It brings together phenomena the most diverse and discovers the hidden analogies which unite them. If matter escapes us, as that of air and light, by its extreme tenuity, if bodies are placed far from us in the immensity of space, if man wishes to know the aspect of the heavens at successive epochs separated by a great number of centuries, if the actions of gravity and of heat are exerted in the interior of the earth at depths which will be always inaccessible, mathematical analysis can yet lay hold of the laws of these phenomena. It makes them present and measurable, and seems to be a faculty of the human mind destined to supplement the shortness of life and the imperfection of the senses; and what is still more remarkable, it follows the same course in the study of all phenomena; it interprets them by the same language, as if to attest to the unity and simplicity of the plan of the universe....

—Joseph Fourier, *The Analytical Theory of Heat*, pp. 7–8

Prologue

Mathematician Michael Frazier of Michigan State University was taught that "real" mathematics by "real" mathematicians is and should be useless. "I never expected to do any applications—I was brought up to believe I should be proud of that," he says. "You did pure harmonic analysis for its own sake, and anything besides that was impure, by definition." But in the summers of 1990 and 1991, he found himself using a mathematical construction to pick out the pop of a submarine hull from surrounding ocean noise.

In St. Louis, Victor Wickerhauser was showing the FBI how the same mathematics might be used to store fingerprints more economically, while at Yale, Ronald Coifman used it to coax a battered recording of Brahms playing the piano into yielding its secrets. In France, Yves Meyer of the University of Paris-Dauphine found himself talking to astronomers about how these new techniques could be used to study the large-scale structure of the universe.

Over the past few years, a number of mathematicians used to the abstractions of pure research have been dirtying their hands—with enthusiasm—on a surprising range of practical projects. What these disparate tasks have in common is a new mathematical language, its alphabet consisting of undulations called wavelets, appropriately stretched, squeezed, or moved about.

A whole range of information—your voice, your fingerprints, a snapshot, x-rays ordered by your doctor, radio signals from outer space, seismic waves—can be translated into this new language, which emerged independently in a number of different fields; in fact, it was only recently understood to be a single language. In many cases this transformation into wavelets makes it easier to transmit, compress, and analyze information or to extract information from surrounding "noise"—even to do faster calculations.

Wavelets quickly won an impressive number of converts. When Olivier Rioul began his doctoral thesis at the Ecole Nationale Supérieure des Télécommunications in Paris in 1989, "wavelets were still a marginal subject in signal processing, as far as the scientific community at large was concerned...." Three years later one could no longer keep count of "the number of researchers publishing in the field,... of theses devoted to it, or of books that have appeared or are about to appear...." [Rioul, pp. 1–2].

In their initial excitement some researchers even thought that wavelets might virtually supplant the much older and very powerful mathematical language of Fourier analysis, used every time you talk on the telephone or turn on a television. But now they see the two as complementary and are exploring ways to combine them, or even to create languages beyond wavelets.

Different languages have different strengths and weaknesses, points out Meyer, one of the founders of the field. "French is effective for analyzing things, for precision, but bad for poetry and conveying emotion—perhaps that's why the French like mathematics so much. I'm told by friends who speak Hebrew that it is much more expressive of poetic images. So if we have information, we need to think, is it best expressed in French? Hebrew? English? The Lapps have 15 different words for snow, so if you wanted to talk about snow, that would be a good choice."

In data processing, some tasks are best tackled with Fourier transforms, others with wavelets; yet other tasks may require new languages. For the first time in a great many years—almost two centuries, if one goes back to the very birth of Fourier analysis—there is a choice.

CHAPTER I

Fourier Analysis:
A Poem Transforms Our World

Although wavelets represent a departure from Fourier analysis, they are also a natural extension of it: the two languages clearly belong to the same family. The history of wavelets thus begins with the history of Fourier analysis. In turn, the roots of Fourier analysis predate Fourier himself (and much of what is now called Fourier analysis is due to his successors). But Fourier is a logical starting point; his influence on mathematics, science, and our daily lives has been incalculable, if to many people invisible. Yet he was not a professional mathematician or scientist; he fit these contributions into an otherwise very busy life.

His father's 12th child, and his mother's ninth, Jean Baptiste Joseph Fourier was born in 1768 in Auxerre, a town roughly halfway between Paris and Dijon. His mother died when he was nine, and his father died the following year. Although two younger siblings were abandoned to a foundling hospital after their mother's death, Fourier continued school and in 1780 entered the Royal Military Academy of Auxerre, where at age 13 he became fascinated by mathematics and took to creeping down at night to a classroom where he studied by candlelight.

"It is said that during the day he secretly collected candle stubs, and that at night, when everyone was asleep, he woke up and crept down to the classroom, lit the candles, and spent long hours working on mathematical problems," related V. Cousin in his *Notes biographiques pour faire suite à l'éloge de M. Fourier,* published in 1831. Fourier's academic success attracted the attention of the local bishop. But when at the end of his studies his application to join the artillery or the army engineers was rejected, he entered the abbey of St. Benoit-sur-Loire. (The popular story that he was rejected by the army because he was not of noble birth—and therefore ineligible "even if he were a second Newton"—is questioned by at least two of Fourier's contemporaries [Cousin, p.2]).

The French Revolution erupted before he took his vows. At first indifferent, he became increasingly committed to the cause of establishing "a free government exempt from kings and priests" [Fourier 2], and in 1793 joined the revolutionary committee of Auxerre. Twice he was arrested, once in the bloody days shortly before the fall of Robespierre and again in June 1795, on charges of terrorism, when he was roused from his bed and scarcely given time to get dressed. As he and his guards left his home, the concierge told Fourier she hoped he would soon be freed; according to Cousin, he never forgot the guard's reply: "You can come get him yourself—in two pieces."

Defending himself, Fourier pointed out that during the Terror no one in Auxerre was condemned to death; "no family in Auxerre," he wrote, "had to mourn a father or a relation" [Fourier 2]. Cousin even relates that once, to prevent a man he believed to be innocent from arrest and the guillotine, Fourier invited the agent charged with the arrest to lunch at an inn, and "having exhausted every means of retaining his guest voluntarily, left the room on a pretext, quietly locked the door, and ran to warn the man threatened by such imminent danger," returning later with excuses.

After the Revolution, Fourier taught in Paris, then accompanied Napoleon to Egypt and served as permanent secretary of the Institute of Egypt. Later he wrote a book on Egypt; even today some people know him as an Egyptologist and are unaware of his contributions to mathematics and physics.

On Fourier's return to France in 1802, Napoleon appointed him prefect of the department of Isère. He served as prefect for 14 years, in Grenoble, earning a reputation as an able administrator; one of his accomplishments was to persuade 37 communes to work together to drain some twenty thousand acres of swamps that had caused annual epidemics of fever, a task that, according to Cousin, required all his tact and "an inexhaustible patience." After Waterloo, denied a government pension because he had served under Napoleon, he found a safe haven in the Bureau of Statistics in Paris, and in 1817 (after an initial rebuff by King Louis XVIII) he was elected to the Academy of Sciences.

A Mathematical Poem

Despite administrative duties and his isolation from Paris for many years, Fourier managed to pursue his scientific and mathematical interests. Victor Hugo called him a man "whom posterity has forgotten" [Hugo], but Fourier's name is as familiar to countless scientists, mathematicians, and engineers as the names of their own children. This fame rests on ideas he set forth in a memoir in 1807 and published in 1822 in his book, *La Théorie Analytique de la Chaleur* (*The Analytic Theory of Heat*).

Physicist James Clark Maxwell called Fourier's book "a great mathematical poem" [Maxwell], but the description does not begin to give an idea of its influence. In the seventeenth century, Isaac Newton had a new insight: that forces are simpler than the motions they cause, and the way to understand the natural world is to use differential and partial differential equations to describe these forces ("perhaps," said Albert Einstein, "the greatest intellectual stride that it has ever been granted to any man to make" [Einstein, p. 68].)

Newton's differential equation showing how the gravitational pull between two objects is determined by their mass and the distance between them replaced countless observations; predictive science became possible, leading the French mathematician and astronomer Pierre Simon Laplace to imagine a single formula that would describe the motions of every object in the universe, for all time.

"An intelligence that knew, for a given instant, all the forces that animate nature and the respective states of everything that composes it, and that in addition was vast enough to submit these facts to analysis, would encompass in the same formula all the movements of the largest bodies of the universe and those of the lightest atom," he wrote, a century after Newton [Laplace, p. vi.]. "Nothing would be uncertain to it, and the future, like the past, would be present before its eyes."

This optimism foundered on the shoals of reality. Solving differential equations—actually predicting where we will be taken by forces that themselves depend at each moment on our changing position—is not easy. As Fourier wrote, although the equations that describe the propagation of heat have a very simple form,

> ... the known methods do not furnish any general mode of integrating them; we could not therefore deduce from them the values of the temperatures after a definite time. The numerical interpretation ... is however necessary.... So long as it is not obtained, the solutions may be said to remain incomplete and useless, and the truth which it is proposed to discover is no less hidden in the formulae of analysis than it was in the physical problem itself. We have applied ourselves with much care to this purpose, and we have been able to overcome the difficulty in all the problems of which we have treated [Fourier 1, p. 21].

Some 150 years after Newton, Fourier provided a practical way to extract the truth from a whole class of such equations: linear partial differential equations.

The reaction of his contemporaries was less enthusiastic than he might have hoped; his memoir was awarded the grand prize for mathematics in 1812, but with the comment that "his analysis ... leaves something to be desired in regards to its generality, and even from the point of view of rigor" [Carslaw, p. 7]. The judgment of posterity has been more generous. The English physicist Lord Kelvin found it "difficult to say whether [Fourier's results'] uniquely original quality, or their transcendently intense mathematical interest, or their perennially important instructiveness for physical science, is most to be praised" [Thomson, p. 578].

His ideas dominated mathematical analysis for a hundred years, having surprising ramifications even for number theory and probability. Outside mathematics their influence is difficult to exaggerate. Virtually every time scientists or engineers model systems or make predictions, they use Fourier analysis. Fourier's ideas are used in linear programming, crystallography, and in countless devices from telephones to radios and hospital x-ray machines. They are, in mathematician T.W. Körner's words [Körner, p. 221], "built into the commonsense of our society."

A Rabble of Functions

There are two parts to Fourier's contribution: first, a mathematical statement (actually proved later by Dirichlet); and second, an explanation of why this statement is useful. The mathematical statement is that any periodic function can be represented as a sum of sines and cosines—what is now called a *Fourier series*. Roughly, what this means is that any curve that periodically repeats itself, no matter how jagged or irregular, can be expressed as the sum of perfectly smooth oscillations (sines and cosines), as shown in Figure 1. The irregular curve and the sum of sines and cosines are two different representations of the same object in different "languages."

The trick is to multiply the sines and cosines by a coefficient to change their amplitude (the height of their waves) and to shift them so that they either add or cancel (changing the phase). Certain nonperiodic functions (those that decrease fast enough so that the area under their graphs is

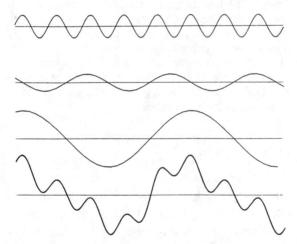

Figure 1. In 1807 Fourier showed that virtually any periodic function can be expressed as the sum of a series of sines and/or cosines. Here the bottom function is composed of the three functions above it. Fourier's realization contributed to a wrenching change in the way mathematicians thought of functions. It provided a straightforward way to solve certain differential equations, and paved the way for digital technology, including computers and compact disks. Fourier analysis is also a natural language for quantum mechanics.

finite) can also be treated this way, using the *Fourier transform*. From the Fourier series, or the Fourier transform, one can reconstruct the original function; no information is lost in translating from one language to the other.

Beyond Plain English 1

The Fourier Transform

The Fourier transform is a mathematical prism, breaking up a function into the frequencies that compose it, as a prism breaks up light into colors. It transforms a function f that depends on time (or on space) into a new function, $\hat{f}$, or "f hat," which depends on frequency. This new function is called the *Fourier transform* of the original function— or, when the original function is periodic, its *Fourier series*.

A function and its Fourier transform are two faces of the same information. The function displays the time (or space) information and hides the information about frequencies. The Fourier transform displays information about frequencies, but information about time or space is hidden in the *phases*: the displacement of the sines and cosines for each frequency, so that they add or subtract. The Fourier transform of music tells what notes (frequencies) are played, but it is virtually impossible to discern when the notes are played.

The Fourier series of a periodic function concerns only those sines and cosines that are integer multiples of the base frequency: for example, $\sin 2\pi k$, $\sin 2\pi 2k$... To make the Fourier transform of a (nonperiodic) function that decreases fast enough at infinity we must compute coefficients of all possible frequencies.

See p. 117.

Fourier himself found this statement "quite extraordinary," and it met with some hostility. Mathematicians were used to functions whose graphs are regular curves; the function $f(x) = x^2$, for example, produces a well-behaved, symmetrical parabola. (A *function* gives a rule for changing an arbitrary number into something else; the function $f(x) = x^2$ says to square any number x: if $x = 2$, then $f(x) = 4$.) The idea that an arbitrary periodic curve could be expressed as a series of sines and cosines and thus treated as a function came as a shock and contributed to a profound and sometimes disturbing change in mathematics. As we shall see in *Traveling from One Function Space to Another*, p. 223, mathematicians spent much of the nineteenth century coming to terms with these changes; eventually they were forced to admit as functions even some mathematical objects that are too wild to be graphed or imagined, but which translate into perfectly civilized Fourier series. "...we have seen a rabble of functions arise whose only job, it seems, is to look as little as possible like decent and useful functions," lamented the French mathematician Henri Poincaré in 1889 [Poincaré 1, pp. 130–131]. "No more continuity, or perhaps continuity but no derivatives.... Yesterday, if a new function was invented it was to serve some practical end; today they are specially invented only to show up the arguments of our fathers, and they will never have any other use."

Poincaré (who also wrote that Fourier's book "was of paramount importance in the history of mathematics and that pure analysis perhaps owed it even more than applied analysis" [Poincaré 2, p. 1]), worried about the effect these bizarre functions would have on a beginning student: "What will the poor student think? He will think that mathematical science is just an arbitrary heap of useless subtleties; either he will turn from it in aversion, or he will treat it like an amusing game...."

Ironically, Poincaré himself was ultimately responsible for showing that seemingly "pathological" functions are essential in describing nature (leading to such fields as chaos and fractals), and this new direction for mathematics proved enormously fruitful, giving new vigor to a discipline that some had found increasingly anemic, if not moribund.

In 1810 the French astronomer Jean-Baptiste Delambre (who had measured, with Pierre Méchain and under most difficult conditions, the arc of meridian between Dunkerque and Barcelona to establish the length of the new meter) had issued a report on mathematics expressing the fear that "the power of our methods is almost exhausted" [Delambre, p. 125]. Some 30 years earlier, Lagrange, then 45 years old, wrote to his friend d'Alembert that he wasn't sure he would be doing mathematics in another ten years. "It seems to me that the mine is already too deep, and that unless we find new veins, sooner or later we will have to abandon it.... It is not impossible that the mathematical positions in the academies will one day become what the university chairs in Arabic are now" [Lagrange].

"Looking back," writes Körner [Körner, p. 474], we can see Fourier's memoir "as heralding the surge of new mathematical methods and results which were to mark the new century."

Beyond Plain English 2

**The Convergence of Fourier Series and the
Stability of the Solar System**

Virtually every periodic function can be represented as a series, or
sum, of sines and cosines, but not every series of sines and cosines
represents a function. If a series can be proved to converge, one can
work with a finite number of terms, confident that adding more terms
won't significantly change results. But if the coefficients of the series
do not become small fast enough, the series diverges and does not
represent a function.
Questions of convergence and divergence have provided a great deal
of work for mathematicians. A particularly difficult problem is determining whether a series with an infinite number of small divisors is convergent. The great nineteenth-century German mathematician Karl
Weierstrass struggled for years in an unsuccessful attempt to prove
the stability of the solar system by proving the convergence of a series
with small divisors. The problem was finally solved in the 1960s.

See p. 125.

The Explanation of Natural Phenomena

The German mathematician Carl Jacobi wrote that Fourier believed the
chief goal of mathematics to be "the public good and the explanation of
natural phenomena" [Jacobi, p. 454]. Fourier showed how his mathematical statement could be used to study natural phenomena such as heat
diffusion, making it possible to numerically solve equations that had until then remained intractable. For a very important class of differential
equations, the Fourier transform turns a difficult differential equation into
a series of simple equations.

Suppose, for example, we want to predict the temperature at time t of
each point along a metal bar that is cooling. As shown in Figure 2, we
start by establishing the initial temperature—at time "zero"—which we
consider as a function of distance along the bar. The differential equation
that describes how the temperature varies doesn't look particularly for-
bidding, but it has two independent variables, time and distance; before
Fourier, no one knew how to tackle it. But when that function is translated

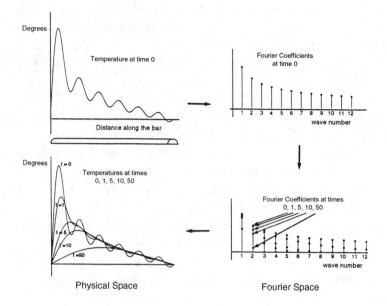

Physical Space   **Fourier Space**

Figure 2. Different Equations in Fourier Space. To determine the temperature at time t of a metal bar that is cooling, one starts by measuring the bar's initial temperature (at time 0), representing it as a function f that depends on space (distance along the bar). Next one goes into Fourier space, calculating its Fourier transform $\hat{f}$, which tells us the coefficient for each wave number making up the function f at time 0. (For a function that depends on time, we would say frequency rather than wave number.) The Fourier coefficients at time 0, $c_n(0)$ (where n is the wave number, 1, 2, 3...), are given by the formula $c_n(0) = \frac{1}{\sqrt{n}}$. The coefficients at time t are computed with the formula

$$c_n(t) = c_n(0)\, e^{n^2 t/100} = \frac{e^{-n^2 t/100}}{\sqrt{n}}$$

We will consider here the coefficients for times $t = 1$, 5, 10, and 50. For each such time, the coefficients are the same for the entire bar: the information on space seems to have disappeared. But it reappears when we return to physical space: when we "invert" (undo) the Fourier transform $\hat{f}_t$, we obtain f_t, which gives the temperature for each point of the bar at time t.

into a Fourier series, a remarkable thing happens: this intractable differential equation *decouples*, becoming a series of independent differential equations, one for each sine or cosine making up the original function.

Each equation tells how the Fourier coefficient of one particular sine or cosine varies with time. (The coefficients, we will recall, are the numbers by which we multiply the sines and cosines of different frequencies to make them taller or shorter; they indicate how much of each frequency a function contains.) The equations are moreover very simple—the same as the equation that gives the value of a bank account earning compound interest (negative interest in this case).

One by one, we simply plug into the solutions of these equations the coefficients describing the temperature at time zero, and then we crank out the answers; these are the Fourier coefficients of the temperature at time t. Now we use these new coefficients to construct a new function giving the new temperature at each point on the bar. The procedure is no harder than the one your bank uses to compute the balance in your account each month.

Essentially we have made a little detour in Fourier space, where our calculations are immensely easier—as if, faced with the problem of multiplying the Roman numerals LXXXVI and XLI, we translated them into Arabic numerals to calculate 86 x 41 = 3526 and then translated the answer back into Roman numerals: LXXXVI $\times$ XLI = MMMDXXVI.

This example shows why Fourier needed a technique that could be applied to any function, even discontinuous; he couldn't expect the initial temperature to be so obliging as to take the form of a regular curve. "In order that these solutions might be general, and have an extent equal to that of the problem, it was requisite that they should accord with the initial state of the temperatures, which is arbitrary," he wrote [Fourier 1, p. 22].

The Public Good

The techniques Fourier invented have had an impact well beyond studies of heat, or even solutions to differential equations. Real data tend to be very irregular. Consider an electrocardiogram, or the readings of a seismograph. Such signals often look like "complicated arabesques," to use Yves Meyer's expression [Meyer 3, p. 2]—tantalizing curves that contain all the information of the signal but hide it from our comprehension. (We will speak of *signals* to be analyzed and of *functions*, such as wavelets, to analyze signals, but mathematically, signals are also functions.)

Fourier analysis translates these signals into a form that makes sense, transforming a signal that varies with time (or in some cases, with space) into a new function, the *Fourier transform* of the signal, which tells how much of each frequency the signal contains (more precisely, how much of the sine and cosine of each frequency it contains).

In many cases these frequencies are not simply a mathematical trick to make calculations easier; they correspond to the frequencies of the actual physical waves making up the signal. When we listen to music or conversation, we hear changes in air pressure caused by sound waves; high sounds have high frequency, with waves close together, and low sounds have lower frequency, the waves more spread out. (A piano can perform a kind of Fourier analysis: a loud sound near a piano with the damper off will cause certain strings to vibrate, corresponding to the different frequencies making up the sound.)

Similarly—although this was not known in Fourier's time—radio waves, microwaves, infrared, visible light, and x-rays are all electromagnetic waves differing only in frequency. Being able to break down sound waves and electromagnetic waves into frequencies has myriad uses, including tuning your radio to your favorite station, interpreting radiation from distant galaxies, using ultrasound to check the health of a developing fetus, and making cheap long-distance telephone calls.

With the discovery of quantum mechanics, it became clear that Fourier analysis is the language of nature itself. On the "position space" side of the Fourier transform, one can talk about an elementary particle's position; on the other side, in "Fourier space," one can talk about its momentum or think of it as a wave. The modern realization that matter at very small scales behaves differently from matter on a human scale—that an elementary particle does not simultaneously have a precise position and a precise momentum—is a natural consequence of Fourier analysis.

The Sampling Theorem and Digital Technology

While irregular functions can be expressed as sums of sines and cosines, usually those sums are infinite. Why translate a complex signal into an endless arithmetic problem in which one must calculate an infinite number of coefficients and sum an infinite number of waves? We seem to be jumping out of the pot into the frying pan. Fortunately a small number of coefficients is often adequate. In the case of the heat diffusion equation, for example, Fourier showed that the coefficients of high-frequency sines and cosines rapidly approach zero, so all but the first few frequencies can safely be ignored. In other cases engineers may assume that a limited number of calculations gives a sufficient approximation, until proved otherwise.

In addition, engineers and scientists using Fourier analysis often don't bother to add up the sines and cosines to reconstruct the signal; instead they "read" Fourier coefficients (or at least the amplitudes; phases are more difficult) to get the information they want, the way some musicians can hear music silently by reading the notes. They may spend hours on end working happily in this "Fourier space," rarely emerging into "physical space." (For one-dimensional signals, "physical space" generally corresponds to time, but Fourier analysis can also be applied to pictures, as shown in Figure 3. In this case, "physical space" corresponds to position.) But the time it takes to calculate Fourier coefficients is a problem: without computers and fast algorithms, Fourier analysis would have remained a theoretical tool, and digital technology would not pervade modern life.

The basis for digital technology was given by the *sampling theorem*, proved by the mathematician J. Whittaker in 1935 [Whittaker] and applied to communication theory in 1949 by Claude Shannon, a mathematician at Bell Laboratories whose contributions to information theory made him a "hero to all communicators" [Pierce, Noll, p. 55]. This theorem proved that if the range of frequencies of a signal measured in cycles per second is n, then the signal can be represented with complete accuracy by measuring its amplitude $2n$ times a second.

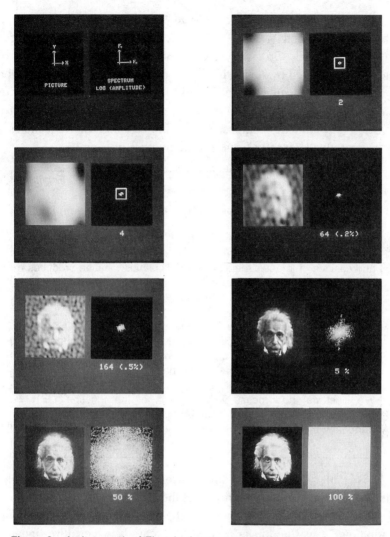

Figure 3. A photograph of Einstein decomposed into its "spatial frequencies." Just as a one-dimensional signal can be represented as a superposition of sines and cosines of different frequencies, a two-dimensional image can be decomposed into different spatial frequencies. But in addition to the amplitude and phase of the sines and cosines, their orientation must be specified. Here we see a photograph reconstructed by successively adding Fourier components, starting with the frequencies that contribute the most to the picture. To the right of each image is the part of the total spectrum represented; the first two spectra have been enlarged. (Courtesy of Gene Switkes, Karen De Valois, and Russell De Valois.)

"This is really a remarkable theorem," comments Warren Weaver, co-author with Shannon of *The Mathematical Theory of Communications*, published in 1949. "Ordinarily a continuous curve can be only approximately characterized by stating any finite number of points through which it passes, and an infinite number would in general be required for complete information about the curve" [Shannon, Weaver, p. 12]. But if the curve is *band-limited*—composed of a limited range of frequencies—it can be reproduced exactly from a finite number of samples.

Beyond Plain English 3

Computing Fourier Coefficients with Integrals

To compute the Fourier coefficients of a periodic function f of period 1, we multiply f by the functions $\sin 2\pi kx$ and $\cos 2\pi kx$ (sines and cosines of integer frequencies k). This multiplication produces a function whose graph oscillates between the graphs of $+f$ and $-f$. The integral of this product (the area delimited by the graph of the new function) is the Fourier coefficient at a given frequency.

At very high frequencies, the Fourier coefficients of a smooth function tend towards zero. The function changes slowly relative to the rapid, high-frequency oscillations, and these oscillations tend to delimit almost equal amounts of negative and positive areas, producing very small coefficients.

We illustrate the process with a graph of the function $f(x) = |\sin(3 \sin 2\pi x)|$, and graphs of that function multiplied by cosines of frequency 7, 8, 30, and 100.

See p. 137.

This result, which follows directly from Fourier analysis, is simple to state and not very difficult to prove (see the Appendix, p. 277, for a proof), but it had enormous consequences for the transmission and processing of information. It was no longer necessary to reproduce the entire signal—to make an *analog*: a limited number of samples suffice.

Since the range of frequencies transmitted by a telephone line is about 4,000 hertz (cycles per second), when you talk on the telephone your voice is measured about 8,000 times a second. Reproducing music with fidelity requires a much broader band of frequencies; when you listen to a compact disc you hear the results of about 44,000 samples a second.

Measuring the signal more often, or reproducing it continuously as with old-fashioned records, does not gain anything.

Another consequence is that high frequencies must be sampled more often than low frequencies, since frequency doubles every time one goes up an octave. The range of frequencies between the lowest two A's on a piano is only 28 hertz, while that between the two highest A's is 1760 hertz—although we hear both spreads as octaves. Encoding a piece of music played in the highest octave would require 3520 samples a second; in the lowest octave, 56 would do.

Beyond Plain English 4

The Fast Fourier Transform

By cutting from n^2 to $n \log n$ the number of computations necessary to compute a Fourier transform, the fast Fourier transform (FFT) transformed entire industries, as well as areas of research—like crystallography—that rely heavily on the Fourier transform.
The idea goes back to Gauss in about 1805, predating Fourier analysis. It can be explained in different ways. Some mathematicians and engineers greatly prefer the traditional approach, but we found it somewhat confusing. We've used a different approach, suggested by Gilbert Strang of MIT, based on a matrix factorization. First we show how to compute a "slow" Fourier transform, using a simple example, and then we lay out the problem in matrix form and show how a clever factorization of matrices "greatly reduces the tediousness of mechanical calculations," as Gauss himself put it in a paper published only after his death.

See p. 141.

The sampling theorem opened the door to digital technology: a sampled signal could be expressed as a series of digits (focusing attention on the problem of round-off errors). With Fourier analysis, your voice can even be shifted temporarily into different frequencies so that it can share the same telephone line with many other voices, contributing to enormous savings. (In 1915 a three-minute call from coast to coast cost more than $260 in today's dollars. [Pierce, Noll, p. 85].)

In 1948, Shannon and his colleagues expected digital transmission to "sweep the field of communications" [Pierce, Noll, p. 79]; the revolution came later than they had expected, but when it came it changed everything. In a few years a top-of-the-line electric typewriter became almost as old-fashioned as a spinning wheel.

Fueling this revolution was the fast Fourier transform (FFT), a mathematical trick that catapulted the calculation of Fourier coefficients out of horse-and-buggy days into supersonic travel. With it, calculations could be done in seconds that previously were too costly to do at all. "It's the difference between being academic and being real," says Michael Frazier of Michigan State University. This fast algorithm requires computers to be useful. "Once the method was established it became clear that it had a long and interesting prehistory going back as far as Gauss. But until the advent of computing machines it was a solution looking for a problem," writes Körner [Körner, p. 499]. On the other hand, the gain in speed from the FFT is greater than the gain in speed from better computers; significant gains in computer speed have come from such fast algorithms built into computer hardware.

CHAPTER II

Seeking New Tools

The fast Fourier transform had a far-reaching impact on society. But it could be argued that it was too successful. "Because the FFT is very effective, people have used it in problems where it is not useful—the way Americans use cars to go half a block," says Yves Meyer of the University of Paris-Dauphine. "Cars are very useful, but that's a misuse of the car. So the FFT has been misused, because it's so practical." Fourier analysis does not work equally well for all kinds of signals or for all kinds of problems. In some cases, scientists using it are like the man looking for a coin under a lamppost, not because that is where he dropped it, but because that's where the light is.

Fourier analysis works with linear problems. Nonlinear problems tend to be much harder, and the behavior of nonlinear systems is much less predictable: a small change in input can cause a big change in output. The law of gravity is nonlinear, and using it to predict the very long-term behavior of even three bodies in space is wildly difficult, perhaps impossible; the system is too unstable. (Engineers make clever use of this instability when sending space probes to distant planets: a probe will pass close enough to the orbit of one planet to pick up momentum, and then a little nudge from booster rockets will break it loose from the planet's gravitational field, enabling it to escape the orbit and proceed on its journey.)

"It is sometimes said," quips Körner [Körner, p. 99], "that the great discovery of the nineteenth century was that the equations of nature were linear, and the great discovery of the twentieth century is that they are not."

21

Engineers faced with a nonlinear problem often resort to the rough-and-ready expedient of pretending that it's linear and hoping for the best. High waves flood Venice every year, forcing Venetians to make their way across the Piazza San Marco on sidewalks set on stilts; engineers would like to be able to predict the flood waters far enough in advance so that they could raise inflatable dikes to protect the city. Since they can't solve the relevant nonlinear partial differential equation (an equation involving winds, position of the moon, atmospheric pressure, and so on), they reduce it to a linear equation, and solve it with Fourier analysis. Despite some progress, they are still taken by surprise by sudden rises in water level of up to a meter.

A Distortion of Reality

Fourier analysis has other limits as well. While mathematically it is "beyond reproach, even experts could not at times conceal an uneasy feeling when it came to the physical interpretation of results obtained by the Fourier method," wrote Dennis Gabor (later to receive the Nobel Prize for the invention of holography) in an article published in 1946 [Gabor, p. 431]. The building blocks of Fourier analysis are sines and cosines, which oscillate for all time. In this framework of infinite time, " 'changing frequency' becomes a contradiction in terms," Gabor wrote. Yet we are all aware of the changing frequencies of sirens, speech, and music.

A Fourier transform hides information about time. It proclaims unambiguously how much of each frequency a signal contains, but is secretive about when these frequencies were emitted. It pretends, so to speak, that any instant of a signal is identical to any other, even if the signal is as complex as a Mozart symphony or changes as dramatically as the electrocardiogram of a fatal heart attack.

The information about time is not destroyed (if it were, we could not reconstruct the signal from the transform), but it is buried deep within the phases. The same sines and cosines can represent very different moments in a signal because they are shifted in phase so that they amplify or cancel each other. Imagine sending waves of different heights and wavelength across a lake, timing the beginning of each wave—shifting their phases—so that the waves would cancel in some places and add in others. Instead of a lake ruffled by uniform waves, this imaginary lake might consist of mirror-smooth pools flanked by 10-foot waves, with regular ripples further on.

For sound and electromagnetic signals this bizarre phenomenon is routine; the same waves, oscillating unaltered, combine to form signals that change constantly. Singly they carry no information. Together, they can transmit both the finale of a symphony and a moment of silence, both the agitated lines of the electrocardiogram of a beating heart and the flat lines that announce death. This phenomenon of constant change built from immutable elements is hard to reconcile with our physical experience and intuition; physicist J. Ville even called it "a distortion of reality" [Meyer 3, p. 63]. The finale of a symphony does not sound like silence; life is not death. So Fourier analysis is poorly suited to very brief signals, or signals that change suddenly and unpredictably; yet in signal processing, brief changes often carry the most interesting information.

In theory one can extract this time information by calculating the phases from the Fourier coefficients. In practice, computing them with enough precision is impossible, and the fact that information about one instant of a signal is dispersed throughout all the frequencies of the entire transform is a serious drawback. A *local* characteristic of the signal becomes a *global* characteristic of the transform. A discontinuity, for example, is represented by a superposition of all possible frequencies. Even inferring from such a superposition that the signal is discontinuous somewhere is not always possible—much less saying where the discontinuity is.

In addition, the lack of time information makes a Fourier transform terribly vulnerable to errors. "If, when you encode an one-hour signal, there is a mistake in the last five minutes, that mistake will corrupt the entire Fourier transform," points out Meyer. The information in one part of a signal, whether real or erroneous, is necessarily spread throughout the entire transform. And errors of phase are disastrous; the least mistake and one may land on something that bears no resemblance to the original signal.

Searching for Lost Time: Windowed Fourier Analysis

While Fourier analysis forces us to choose between time on one side of the transform and frequency on the other, ". . . our everyday experiences—especially our auditory sensations—insist on a description in terms of both time and frequency," Gabor wrote [Gabor, p. 429]. To analyze a signal in both time and frequency, he used the windowed Fourier transform.[1] The idea is to study the frequencies of a signal, segment by segment; that way, one can at least limit the span of time during which something is happening. The "window" that defines the size of the segment to be analyzed—and which remains fixed in size—is a little piece of curve; that curve is successively filled with oscillating functions of different frequencies. (See Figure 2, page 29.)

While the classical Fourier transform compares the entire signal successively to infinite sines and cosines of different frequencies, to see how much of each frequency it contains, windowed (or "short-time") Fourier analysis compares a segment of the signal to bits of oscillating curves, first of one frequency, then of another, and so on. When one segment of the signal has been analyzed, one "slides" the window along the signal, to analyze another segment.

But this method imposes painful compromises. The smaller your window, the better you can locate sudden changes, such as peaks or discontinuities—but the blinder you become to the lower frequency components of your signal. These lower frequencies just won't fit into the little window. If you choose a bigger window, you can see more of the low frequencies, but you do worse at "localizing in time."

So Yves Meyer, well aware of the power and limitations of Fourier analysis (the fundamental tool of his subject, harmonic analysis), was intrigued when he heard of little waves—*ondelettes* in French—that made

---

[1]Another approach to the time-frequency dilemma, proposed by J. Ville in 1948, was to define a time-frequency "energy density"; the same mathematics, in the context of quantum mechanics, had been introduced by 1932 by E. P. Wigner. The Wigner-Ville *distribution* looks at first glance like an ideal way to analyze a signal simultaneously in time and frequency, but it results in interferences, which can be removed . . . at the cost of decreased resolution. For a discussion of how the Wigner-Ville distribution relates to windowed Fourier analysis and wavelets, see for example Chapter 4 of [Mallat 4].

it possible to decompose signals simultaneously by time and by frequency. Would wavelets give the equivalent of a musical score for each signal, telling not just what notes (what frequencies) to play but also when to play them?

Talking to Heathens

Meyer got involved "almost by accident," he recalls.

I was a professor at the Ecole Polytechnique, where we shared the same photocopy machine with the department of theoretical physics. The department chairman liked to read everything, know everything; he was constantly making photocopies. Instead of being exasperated when I had to wait, I would chat with him while he made his copies. One day in the spring of 1985 he showed me an article by a physicist colleague of his, Alex Grossmann in Marseille, and asked whether it interested me. It involved signal processing, using a mathematical technique I was familiar with. I took the train to Marseille and started working with Grossmann.

Often pure math "trickles down" to applications, but this was not the case for wavelets, Meyer adds. "This is not something imposed by the mathematicians; it came from engineering. I recognized familiar mathematics, but the scientific movement was from application to theory. The mathematicians did a little cleaning, gave it more structure, more order."

Structure and order were needed; the predecessors of today's wavelets had grown in a topsy-turvy fashion, to the extent that in the early days wavelet researchers often found themselves unwittingly recreating work of the past. (According to some, this still happens. "It seems that a lot of people in signal processing should be taught about what we've done in vision, and then go on from there," says David J. Field, a psychologist at Cornell University, voicing a complaint probably felt by people in many fields: people in other fields don't know enough about the work done in theirs.) All these researchers didn't realize they were speaking the same language: partly for the simple reason that they rarely spoke to one another, but also because the early work existed in such disparate forms. Grossmann had spoken about wavelets to other people in Meyer's field, but they "didn't make the connection," he said. "With Yves it was immediate; he realized what was happening."

Wavelet researchers sometimes joke that the main benefit of wavelets is to allow them to have wavelet conferences; behind that joke lies the reality that the modern, coherent language of wavelets has provided an unusual opportunity for people from different fields to speak and work together. "Under normal circumstances the fields are pretty much watertight one to the other," Grossmann says. "So one of the main reasons that many people find this field very interesting is that they have found themselves outside of their usual universe, talking to heathens of various kinds. Anybody who is not in one's little village is a heathen by definition, and people are always surprised to see—'look, they have two ears and a single nose, just like us!' That has been a very pleasant experience for everyone."

"This Must Be Wrong": Morlet's Wavelets of Constant Shape

Tracing the history of wavelets is almost a job for an archaeologist. "I have found at least 15 distinct roots of the theory, some going back to the 1930s," Meyer said. "David Marr, who worked on artificial vision and robotics at MIT, had similar ideas. The physics community was intuitively aware of wavelets dating back to a paper on renormalization by Kenneth Wilson, the Nobel laureate, in 1971."

In mathematics, Littlewood and Paley developed wavelet-like techniques that were applied to the study of trigonometric series. The theory they developed was extended to higher dimensions by Elias Stein, whose work provided a basis for more recent work on wavelet characterization of function spaces. The Argentinian mathematician Alberto Calderón developed a continuous version of wavelets (now known as the Calderón formula) in his studies of complex interpolation and atomic decompositions of Hardy spaces.

In signal processing, work on techniques later understood to be closely linked to wavelets began in 1976, when three French researchers,

A. Croisier, D. Esteban, and C. Galand, introduced a bank of filters that could be used to decompose, subsample and reconstruct a signal [Croisier et al.]. A decade later F. Mintzer [Mintzer] at IBM and M. Smith and T. Barnwell [Smith, Barnwell] at Georgia Institute of Technology independently constructed filters that were later seen to be related to orthogonal wavelet bases.

Yet other researchers developed wavelets—which they called "self-similar Gabor functions"—to model the visual system.

But we can take as a starting point the work of Jean Morlet, a geophysicist with the French oil company Elf-Aquitaine, who developed wavelets as a tool for oil prospecting. (They were never used for that. "There were a few little trials, and then it stopped. Some people were against it, others were for, and then there wasn't any money ..." says Morlet, now retired.)

The standard way to look for underground oil, introduced in the 1960s, is to send vibrations or impulsions under ground and to analyze their echos (direct reflections or backscattering). This analysis is supposed to tell how deep and how thick the various layers are and what they are made of. Roughly speaking, the frequencies of the echos are linked to the thickness of the various underground layers, with high frequencies corresponding to thin layers. "There are hundreds of layers," says Morlet. "All these reflected signals corresponding to the different layers interfere with each other. It's an awful mess. That's what one tries to separate out ..."

To tease information out of this tangle of echos, Fourier analysis was used on the entire signal. As more powerful computers became available, big "windows" were placed here and there on the signal, then, as the price of computing dropped further, windows were placed closer and closer together, even overlapping. "But we got to a point where no matter what one did, it didn't get any better," Morlet says. "We wanted a finer local definition; in particular, we wanted to have access to information about layers of different thicknesses."

To do this, around 1975 Morlet drew on the time-frequency representation of signals by Gabor some 30 years earlier, using windowed Fourier analysis. This system had the disadvantage of being imprecise about time in the high frequencies (unless one made the window very small, which

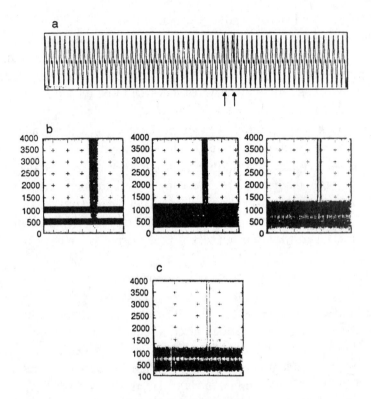

Figure 1. A signal with two pulses analyzed with both windowed Fourier analysis and wavelets. Wavelets give better resolution of the two pulses than even the smallest window. (a) The signal, sampled 8,000 times a second, with pulses indicated by arrows. (b) Three windowed Fourier transforms of the signal, with different window widths: 12.8, 6.4 and 3.2 milliseconds respectively (time varying horizontally, frequency vertically). The widest window gives the best resolution of the two pure tones, but doesn't distinguish between the two pulses. (c) A wavelet transform of the signal. The resolution of the two pulses is even better than with the smallest window in (b), while the resolution of the two tones is comparable to that achieved with the middle window. (Courtesy of Ingrid Daubechies and SIAM.)

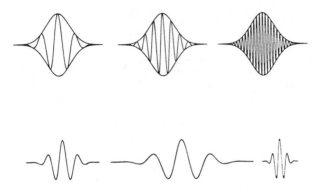

Figure 2. Windowed Fourier Analysis versus Wavelets. Above: In windowed Fourier analysis, the size of the window is fixed and the number of oscillations varies. A small window is "blind" to low frequencies, which are too large for the window. But if one uses a large window, information about a brief change will be lost in the information concerning the entire interval corresponding to the window. Below: A "mother wavelet" (left) is stretched or compressed to change the size of the window. This makes it possible to analyze a signal at different scales. The wavelet transform is sometimes called a "mathematical microscope": big wavelets give an approximate image of the signal, while smaller and smaller wavelets zoom in on small details.

meant losing all information about low frequencies; see Figure 1). It had another serious drawback as well: unlike classical Fourier analysis, it gave no numerical way to reconstruct a signal from the transform.

So Morlet took another tack. Instead of keeping the size of the window fixed and filling it with oscillations of different frequencies, he did the reverse: he kept the number of oscillations in the window constant and varied the width of the window, stretching or compressing it like an accordion or a child's slinky. When he stretched the wavelet, the oscillations inside were stretched, decreasing their "frequency"; when he squeezed the wavelet, the oscillations inside were squeezed, producing higher frequencies (see Figure 2).

Since these new functions had more or less the same shape, whether they were squeezed or stretched, he named them "wavelets of constant shape" to distinguish them both from the Gabor functions (which he called "Gabor wavelets") and from other "wavelets" in geophysics: the signals that are sent underground. (Later, some geophysicists would be surprised at the fuss made about "wavelets"; they knew what wavelets were.)

Making models on a little office calculator, Morlet developed empirical methods for decomposing a signal into wavelets and reconstructing it again. But when he showed his results to others in the field, he was told "this must be wrong, because if it were right, we would know about it." Then in 1981 Morlet asked Roger Balian, a physicist who had been a classmate at Ecole Polytechnique, to help him correct his first paper on wavelets. "Balian told me, I'm a specialist on time-frequency interpretations but I know someone who is more of an expert than I am, and he sent me to see Alex Grossmann in Marseille."

The Error is Zero

"Jean was sent to me because I work in phase space quantum mechanics," Grossmann says. "Both in quantum mechanics and in signal processing you use the Fourier transform all the time—but then somehow you have to keep in mind what happens on both sides of the transform.... When Jean arrived, he had a recipe, and the recipe worked. But whether these numerical things were true in general, whether they were approximations, under what conditions they held, none of this was clear." The two spent a year answering those questions. Their work involved a lot of experimenting on personal computers, Grossmann says:

> One of the many reasons why the whole thing didn't come out earlier is that just about this time it became possible for people who didn't spend their lives in computing to get a little personal computer and play with it. When you try to do something new, try to understand something, you go to something you can handle yourself. Jean did most of his work on a personal computer. Of course he could also handle huge computers, that's his profession, but it's a completely different way of working. And I don't think I could have done anything if I hadn't had a little computer and some graphics output.

To validate their empirical results mathematically, Grossmann and Morlet showed that when wavelets are used to represent a signal, the "energy" of the signal is unchanged. (This energy—the average value of the square of the signal—doesn't necessarily correspond to physical energy.) This means that one can transform a signal into wavelet form and then get exactly the same signal back again—a crucial condition. It also means that a small change in the wavelet representation produces a correspondingly small change in the signal; a little error or change will not be blown out of proportion.

The method of reconstruction was cumbersome, however. Unlike the Fourier transform, which turns a signal with one variable (time or space) into another function with one variable (frequency), the wavelet transform produces a transform with two variables, time and frequency. Reconstructing the signal then required a double integral and was rather painful. But Morlet and Grossmann knew—and had mentioned in several articles—that they could also "approximately" reconstruct the signal using a single integral.

"For practical applications, the price isn't at all the same: one is practically free, and the other, you've got to pay for it," Morlet says. So it would have been interesting to know to what extent such a reconstruction was approximate: was the error big? But Morlet and Grossmann weren't mathematicians and hesitated to tackle the problem. "After a while we said to each other, we really ought to see what we can do with this," Morlet recalls. "And Alex said, all right, I'll try to calculate the error that we make when we reconstruct like that. One day in September 1984 he called me—I was in San Diego and he was in Pasadena—and he told me, I found the error. It's zero. ... "[2]

Wavelets: A Mathematical Microscope

Wavelets are an extension of Fourier analysis. As with the Fourier transform, the point of wavelets is not the wavelets themselves; they are a means to an end. The goal is to turn the information of a signal into numbers—coefficients—that can be manipulated, stored, transmitted, analyzed, or used to reconstruct the original signal.

---

[2]Later they learned that this result had been proved in 1964 in the context of harmonic analysis by the mathematician A. Calderón [Calderón]—who, however, interpreted the result very differently.

The basic approach is the same. The coefficients tell in what way the analyzing function (sines and cosines, or wavelets) needs to be modified in order to reconstruct a signal. One can literally construct the signal by adding together wavelets of different sizes, at different positions, just as one can construct a signal by adding together sines and cosines. The technique underlying the computation of coefficients is the same: the signal and the analyzing function are multiplied together, and the integral of the product is computed. (For some graphs illustrating this process in Fourier analysis, see *Computing Fourier Coefficients with Integrals*, p. 137. In practice, different fast algorithms are used). For the envelope of his wavelets, Morlet even used the Gaussian, or bell-shaped, function used by Gabor in windowed Fourier analysis.

But squeezing and stretching the wavelets to change their frequency changed everything. Wavelets automatically adapt to the different components of a signal, using a small window to look at brief, high-frequency components and a large window to look at long-lived, low-frequency components. The procedure is called *multiresolution*; the signal is studied at a coarse resolution to get an overall picture and at higher and higher resolutions to see increasingly fine details. Wavelets have been called a "mathematical microscope"; compressing wavelets increases the magnification of this microscope, enabling you to take a closer look at small details in the signal. (In searching for oil, looking at higher frequencies corresponds to inspecting thinner layers.)

Typically, as shown in Figure 3, five different resolutions are used, each twice as fine as the previous resolution. (One can also study the signal at intermediate resolutions; with the first wavelets, one had to.) Sometimes people talk about scale, rather than resolution, and Morlet talked about "octaves." Resolution evokes the number of wavelets used— the number of times the signal is sampled. Scale evokes the fact that changing the size of the wavelet changes the size of the components one can see. Octaves evoke the fact that going from one resolution to one twice as fine doubles the frequency of the wavelets so that they encode frequencies that are twice as high. (But the frequency information is approximate. Unlike the sines and cosines of Fourier analysis, a wavelet doesn't have a precise frequency.) The peculiar nature of wavelets—the

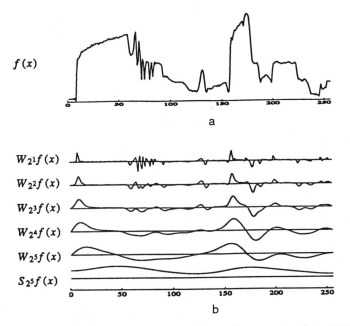

$f(x)$

a

$W_{2^1}f(x)$

$W_{2^2}f(x)$

$W_{2^3}f(x)$

$W_{2^4}f(x)$

$W_{2^5}f(x)$

$S_{2^5}f(x)$

b

Figure 3. An Example of a Wavelet Transform. (a) The original signal. (b) The wavelet transform of the signal, over 5 scales, differing by a factor of 2. The finest resolution, giving the smallest details, is at the top. At the bottom is the graph of the remaining low frequencies. (Courtesy of Stéphane Mallat.)

result of their "constant shape"—is that resolution, scale and frequency all change at once.

Together, all the coefficients at all the scales give a good picture of the signal or function. As Meyer writes [Meyer 2, p. xii], "...contrary to what happens with Fourier series, the coefficients of the wavelet series translate the properties of the function or distribution simply, precisely and faithfully"—at least, he adds today, "the properties that correspond to strong transients: everything that is rupture, discontinuity, the unforeseen." Not just because wavelets make it possible to look closely at the details of a signal, but also because they encode only changes. As shown in Figure 4, a wavelet coefficient measures the correlation, or agreement, between the wavelet (with its peaks and valleys) and the corresponding segment of the signal. "You play with the width of the wavelet in order to catch the rhythm of the signal," Meyer says. "Strong correlation means that there is a little piece of the signal that looks like the wavelet."

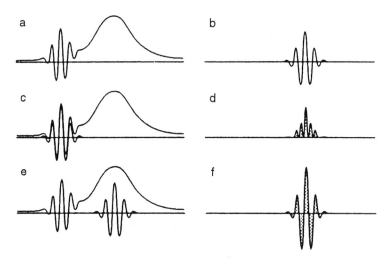

Figure 4. In the wavelet transform, a wavelet (b) is compared successively to different sections of a function (a). The product of the section and the wavelet is a new function; the area delimited by that function is the wavelet coefficient. In (c) the wavelet is compared to a section of the function that looks like the wavelet. The product of the two is always positive, giving the big coefficient shown in (d). (The product of two negative functions is positive.) In (e) the wavelet is compared to a slowly changing section of the function, giving the small coefficients shown in (f). The signal is analyzed at different scales, using wavelets of different widths. "You play with the width of the wavelet in order to catch the rhythm of the signal," says Yves Meyer.

Constant stretches give wavelet coefficients with the value zero. By definition, a wavelet has an integral of zero—half the area it surrounds is positive, the other half negative. Multiplying the wavelet by a constant changes the positive and negative parts equally, and the integral remains zero. One can also construct wavelets that give small or zero coefficients when they are compared to linear functions, quadratics, and even higher degree polynomials. The number of *vanishing moments* determines what the wavelet "doesn't see."

"Wavelet analysis is a way of saying that one is sensitive to change," says Meyer. "It's like our response to speed. The human body is only sensitive to accelerations, not to speed." As long as the speed of a train or airplane is constant, we feel that it isn't moving. This characteristic enables wavelets to compress information. Typically, a signal containing some 100,000 values can be reduced to 10,000 wavelet coefficients; those that are zero are automatically set aside. As Figures 3 and 4 illustrate, it also makes them good at "seeing" changes—peaks or pulses in a signal, for example, or edges in a picture. Wavelet analysis is, in Meyer's words [Meyer 5, p. 5], "an intelligent reading" of these kinds of signals, "jumping straight to essentials."

Tacitus vs. Cicero: The Search for Orthogonality

When Meyer took the train to Marseille to see Grossmann in 1985, the idea of multiresolution existed, but calculating wavelet coefficients was slow and rather painful; in addition, the transformation was not concise. If one wanted to reconstruct the signal later, it was necessary to study it not just at resolutions differing by a factor of two but also at intermediate resolutions. Morlet and Grossmann had devised a system, shown in Figure 5, that used just a few "voices" between the "octaves," and which produced only small errors when the signal was reconstructed.

Beyond Plain English 5

The Continuous Wavelet Transform

In the continuous wavelet transform, a function ψ ("psi") is used to create a family of wavelets $\psi(at + b)$ where a and b are real numbers, a compressing or stretching the function ψ and b displacing it. The continuous wavelet transform turns a signal $f(t)$ into a function $c(a, b)$ with two variables, scale and time:

$$c(a, b) = \int f(t)\, \psi(at + b)\, dt.$$

This transformation is in theory infinitely redundant, but ways have been found to rapidly extract the essential information from these redundant transforms. One method reduces a transform to its *skeleton*; another involves computing the transform's maximum wavelet coefficients, called *wavelet maxima*.

See p. 149.

But if one insisted on perfect reconstruction, it was necessary to use the *continuous* wavelet transform, studying the signal at all possible resolutions and displacing the wavelets by all values, not just discrete values.

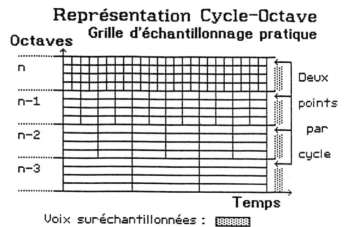

Figure 5. Wavelets and the Sampling Theorem. In wavelet analysis, each wavelet coefficient is equivalent to a sampling of the signal. With his wavelets of constant shape, Morlet doubled the number of samples when he went from one octave to the next higher octave. (Today one would say, "from one scale to the scale twice as small," or "from one resolution to the resolution twice as fine.") He also took samples between the octaves.

These intermediary *voices* meant that the signal was *over-sampled* in terms of the Shannon sampling theorem. With orthogonal wavelets it became possible to use the minimum number of samples required by the sampling theorem—*critical sampling*—and still reconstruct the signal perfectly. (Courtesy of Jean Morlet.)

Imagine a wavelet slowly gliding along the signal while you calculate an infinite number of wavelet coefficients. Once this endless job is finished, you compress the wavelet infinitesimally and start over, calculating another infinite number of coefficients, and so on.... You study the signal not just at resolutions very close to each other but also at the resolutions in between, as well as the resolutions between those resolutions.

In theory, the task is infinite; in practice, "infinite may mean 10,000 coefficients, which isn't so bad," Grossmann says. When people talk about the continuous wavelet transform, most often they are really working with discrete numbers, sampling the signal a finite number of times. But the continuous transform is still extravagant. The wavelets overlap each other, so most of the information encoded by one is also encoded by its neighbors. (The signal is said to be *over-sampled*; there are more samples than is required by the sampling theorem.) Typically, Meyer says, the transform is redundant by a factor of ten: "A continuous transform is Ciceronian, where everything is said ten times."

This repetition can be an advantage. A continuous representation is shift-invariant (mathematicians say, *translation-invariant*). Exactly where on the signal one starts the encoding doesn't matter; shifting over a little doesn't change the coefficients. It follows that it is often easier to analyze data, or recognize patterns, with a continuous transform.

Nor is it necessary to know the coefficients with precision. "It's like drawing a map," says Ingrid Daubechies, professor of mathematics at Princeton University, who has worked with wavelets since 1985.

Most men draw these little lines and if you miss one detail you can't find your way. Women tend to put lots of detail—a gas station here, a grocery store there, lots and lots of redundancy. Suppose you took a bad photocopy of that map; if you had all that redundancy you still could use it. You might not be able to read the brand of gasoline on the third corner but it would still have enough information. In that sense you can exploit redundancy: with less precision on everything you know, you still have exact, precise reconstruction.

But if the goal is to compress information in order to store or transmit it more cheaply, redundancy can be costly. For those purposes one may want an *orthogonal* transform, which gives perfect reconstruction of the original signal while avoiding redundancy (see *Orthogonality and Scalar Products*, p. 153). In an orthogonal transform, such as Fourier analysis, the information of the signal is encoded only once: "Tacitus," Meyer says, "compared to Cicero."

At the time, though, Meyer wasn't thinking in terms of applications like the compression of information; as a self-respecting pure mathematician, he was immersed in the mathematics of wavelets. As far as the continuous transform is concerned, practically any function can call itself a wavelet, as long as it has a zero integral. This is not at all the case for orthogonal wavelets; knowing whether they existed at all was an interesting question (aside from the discontinuous Haar wavelet; see *Multiresolution*, p. 165).

Beyond Plain English 6

Orthogonality and Scalar Products

An orthogonal transform is concise. It allows for perfect reconstruction of the original signal, and is relatively easy to compute: each coefficient is computed with a single *scalar product* of the signal and the analyzing function (in our case, the wavelet). These properties reflect a geometrical relationship. A wavelet cannot be orthogonal in isolation; a wavelet *basis* is orthogonal if each wavelet is perpendicular to all the others.

Each wavelet basis contains an infinite number of wavelets, formed by stretching or compressing a "mother" wavelet. To think of them all as mutually perpendicular we think of each wavelet as a single vector in an infinite-dimensional space. The concept of a scalar product of two vectors then allows us to extend some familiar notions from two- and three-dimensional space (angle, length) to multidimensional spaces. Rather than try to imagine these strange spaces, we work by analogy. This approach gives new power to some elementary high-school mathematics.

See p. 153.

In 1981 Roger Balian had proved that one can't have an orthogonal representation with windowed Fourier analysis using a Gaussian window, or, more generally, any reasonably regular and well-localized window.[3] Meyer was convinced that orthogonal wavelets did not exist either—more precisely, very smooth (infinitely differentiable) orthogonal wavelets that soon get close to zero on either side but never actually reach it. He set out to prove it—and failed, in the summer of 1985, by constructing precisely the kind of wavelet he had thought didn't exist. (He learned later that a Swedish mathematician, J. O. Strömberg [Strömberg] had created other orthogonal wavelets, not as smooth, four years earlier.)

With these new wavelets it became possible to make an economical wavelet transform: a transform containing the same number of points as the signal itself.

---

[3]Balian's proof was incomplete; Ronald Coifman and S. Semmes filled the hole, and then Guy Battle [Battle] found a very different proof. See [Meyer 3] or [Daubechies, p. 108].

A New Language
Acquires a Grammar

The following year, in the fall of 1986, Meyer was giving a course on wavelets in the United States when he received several telephone calls from a persistent 23-year-old graduate student in computer vision at the University of Pennsylvania in Philadelphia. Stéphane Mallat is French, and had been a student at the Ecole Polytechnique, one of France's prestigious *grandes écoles*, when Meyer taught there, but the two hadn't met.

Mallat, now at Ecole Polytechnique and the Courant Institute of Mathematical Sciences at New York University, had heard about Meyer's work on wavelets from a friend while he was vacationing in St. Tropez. To him it sounded suspiciously familiar; on reading the article Meyer had written about his orthogonal wavelets, it occurred to Mallat that this mathematics could be applied to image processing. "I am stupid enough to always think that the idea I have at the moment is going to work out, and this is really lucky," he says now.

It saves me a lot of anguish, but more important, by the time I realize that it does not work, it has led to another idea which of course is going to work. At the end of this recursive process there is indeed something which works, which generally has nothing to do with the original idea, but who cares? So I am always trying to explain to students that they should generate enough enthusiasm to persuade themselves that it will work and it will indeed, at some point.

This time the original idea not only worked but led him further than he could have foreseen. On returning to the United States he called Meyer, who agreed to meet him at the University of Chicago. The two spent three days holed up in a borrowed office—"I kept telling Mallat that he absolutely had to go to the Art Institute in Chicago, but we never had time," Meyer says—while Mallat explained that the multiresolution Meyer and others were doing with wavelets was the same thing that electrical engineers and people in image processing were doing under other names.

"This was a completely new idea," Meyer says. "The mathematicians were in their corner, the electrical engineers were in theirs, the people in vision research like David Marr were in another corner, and the fact that a young man who was then 23 years old was capable of saying, you are all doing the same thing, you have to look at that from a broader perspective—you expect that from someone older."

In three days, the two solved the mathematical difficulties; since Meyer was already a full professor, at his insistence the resulting paper, "Multiresolution approximation and wavelets" [Mallat 1], appeared under Mallat's name alone. The paper made it clear that work that existed in many different guises and under many different names—wavelets, the pyramid algorithms used in image processing, the subband coding of signal processing, the quadrature mirror filters of digital speech processing— were at heart all the same. Using wavelets to look at a signal at different resolutions can be seen as applying a succession of filters: filtering out everything but low frequencies with large wavelets, and filtering out everything but high frequencies with small wavelets.

The realization benefited everyone. Multiresolution theory provided a bridge between the digital world of signal processing and the continuous world of the original wavelet transform, wavelets gaining speed and the world of signal processing gaining new insights into existing techniques. The digital filters of signal processing are essential for fast computations, but it is not always clear to what these computations correspond. "People in signal processing knew how to build the filters but did not understand what was really happening when these filters were cascaded," Mallat says. Wavelets provide an intuitive picture: a signal is decomposed into its components at different scales. Wavelets also have a solid mathematical underpinning that makes it easier to relate decomposition coefficients to characteristics of the signal.

"All those existing techniques were tricks that had been cobbled together, they had been made to work in particular cases, but there was no assurance that they were well founded mathematically. There was no general theory," says Marie Farge of the Ecole Normale Supérieure in Paris, who uses wavelets to study turbulence. (Daubechies would qualify this judgment, which she says "engineers will feel as an insult. They had a nice mathematical theory for what they were doing; they just had no idea it tied in with a bigger functional analysis framework.")

A whole mathematical literature on wavelets existed by 1986, some of it developed before the word *wavelet* was even coined; this mathematics could now be applied to other fields. "Mallat helped people in the quadrature mirror filters community, for example, realize that what they were doing was much more profound and much more general than ad hoc techniques," Farge said. "He showed them that you had theorems, and could do a lot of sophisticated mathematics." In addition, wavelets brought two new concepts—regularity and vanishing moments.

Wavelets benefited in two important ways. First, Mallat showed how to use a new function, the *scaling function*, to compute wavelet coefficients fast—transforming wavelets from an interesting variation on Fourier analysis into a powerful practical tool. As Ronald Coifman of Yale University said:

I view wavelet analysis as a natural extension of traditional Fourier analysis, and therefore on a scientific level a translation of mathematical tools and methods which have been in wide use in mathematical analysis and other sciences for the last 50 years. The main reason for the current flurry of activity comes from our ability to translate these mostly theoretical ideas into working engineering tools through fast computational algorithms like Mallat's. The situation is somewhat analogous to the effect of the FFT on applied areas.

Mallat also gave a systematic way to construct new orthogonal wavelets. More than a recipe, he gave an explanation. Meyer's orthogonal wavelets had emerged almost miraculously from his computations. ("The existence of algorithms with the properties we just described seems to be an accident," he wrote [Meyer 1, p. 212].) But mathematicians don't like miracles, even those certified by a proof in good standing. They want to understand. With the theory of multiresolution, wavelets became comprehensible, even natural. "I found my wavelets by trial and error; there was no underlying concept," says Meyer. "Mallat's paper gave a philosophy, a framework, almost a geometrical idea. In some sense he founded the field in writing that article on multiresolution."

Beyond Plain English 7

Multiresolution

Coming from different directions—pure mathematics and computer vision—Stéphane Mallat and Yves Meyer created multiresolution theory. One result of their work was the fast wavelet transform (see p. 183). Another result was a mathematical theory of orthogonal wavelets. This theory defines four mathematical conditions which, if met, make it possible to view a signal at resolutions differing by a factor of two, and to encode the difference of information from one resolution to the next as orthogonal wavelet coefficients.

The theory also gives a recipe for creating new orthogonal wavelet bases. Both scaling functions and wavelets can be created using a function that is the Fourier transform of a low-pass filter. (In geometrical terms, this function parametrizes a curve on a sphere in four-dimensional space.) Curiously, the bottom line of multiresolution theory is that to compute the wavelet transform of a signal we need neither scaling functions nor wavelets: just very simple digital filters.

See p. 165.

Mother or Amoeba?

With an orthogonal wavelet transform, a signal is analyzed at scales varying always by a factor of two (obeying, without excess, the Shannon sampling theorem, since each time the frequency doubles, one doubles the number of wavelets used to sample the signal). A whole family of little waves is used: what the French call an *ondelette mère* (mother

Figure 1. Mathematicians use the word "dilation" to refer both to expansion and to compression; a wavelet is "dilated"—stretched or compressed—to change the scale at which it analyzes a signal. Large wavelets encode the general trend, while small wavelets give the details. Long before wavelets, the English mathematician Charles Dodgson, also known as Lewis Carroll, explored the effects of dilation in *Alice's Adventures in Wonderland*. "...being so many different sizes in a day is very confusing," Alice complained. (Courtesy of Eleanor Hubbard.)

wavelet), a *fonction père* (the father, or scaling, function), and a great many baby wavelets, some small, some twice as big, and some twice as big again ... But this French terminology "shows a scandalous misunderstanding of human reproduction," objects Cornell mathematician Robert Strichartz [Strichartz, p. 540]. "In fact the generation of wavelets more closely resembles the reproductive life style of an amoeba."

The father—who isn't even necessary in a continuous wavelet transform—plays only an indirect role in the creation of the children. As for the children, they are, as we have seen, clones of the mother, stretched or compressed (in mathematical jargon they are *dilated*, a word used, in defiance of dictionaries, even for contractions). The new wavelets are then shifted, or *translated*, to correspond to different parts of the signal. But the father, or scaling, function plays two important roles. It gives the starting point for the analysis, and it makes it possible to compute wavelet coefficients fast.

The first role comes up when one starts with a physical or analog signal, rather than with samples. "The very first operation is to make the signal digital," says Meyer. One could simply sample it, but then one risks falling on unrepresentative values. Instead one divides the signal into segments, each corresponding to a scaling function, and calculates a coefficient for each. This gives average values.

The size of the scaling function determines the finest resolution used for the analysis. If you were looking at changes in temperature, you might be interested in changes between day and night, or on the scale of a month, a year, a decade, a century, and so on. The scaling function gives the starting point—the smallest scale with which you are going to work.

The Fast Wavelet Transform

For the scaling function's second role, Mallat was inspired by the pyramid algorithms described in the early 1980s by two researchers in image processing, Peter Burt, now with the David Sarnoff Research Center in Princeton, and Edward Adelson, now at MIT. If speed isn't an issue, one can always decompose a signal in wavelets by comparing the signal, at each scale or resolution, to wavelets of the appropriate size. But starting each time with the original signal is slow. It's as if one were to study not just cities and highways, but also streams and back roads, in order to make any map of the United States, whether a map of the entire country that shows only major cities and highways, a more detailed map of New York State, or a Geological Survey map showing every stream and hill in a small region. It makes sense to take advantage of work that has already been done; cursory maps are made from more detailed maps. So although one could start analyzing a signal at the coarsest resolution, when speed matters, one starts with the finest resolution: *fine to coarse.*

The first step is to divide the signal into two parts: a somewhat fuzzy image of the signal, and little details (the general tendency of the signal and the fluctuations). The fuzzy, or smoothed, image is the signal seen at half resolution (with half as many samples). It is created with the help of a low-pass filter associated with the scaling function; that is why the scaling function is sometimes called the smoothing function. The details are the finishing touches that would have to be added to the smoothed signal to reproduce the original signal. They are obtained using a high-pass filter associated with the smallest wavelets.

One can see that there must be a precise mathematical relationship between the scaling function and the wavelets used. If a filter associated with the "father" of another family of wavelets is used, there is no reason to expect that the signal smoothed by the father plus the details encoded by the mother would equal all the information of the signal. The mother, father, and baby wavelets have a striking family resemblance, evoked by this terminology, which has somewhat fallen into disfavor.

Beyond Plain English 8

The Fast Wavelet Transform

The fast wavelet transform (FWT) is a simple and fast method for decomposing a signal into components differing in size by a factor of two. The resulting transform has the same number of points as the original signal, and the process is linear; computing a wavelet transform with the FWT takes cn computations, where c depends on the complexity of the wavelet. For long signals this is faster than the FFT, although the comparison can be misleading, since in practice a long signal is broken into parts before the FFT is applied.

A wavelet coefficient (or scaling function coefficient) can be obtained by integrating the product of the wavelet (or scaling function) with the signal. In practice one does something much simpler, *convolving* the signal with both a short high-pass digital filter associated with the wavelet and a short low-pass digital filter associated with the scaling function. A simple example involving the Haar scaling function and wavelet (whose filters each consist of two digits) shows that the two processes give the same result.

See p. 183.

In the second step, we save the first wavelet coefficients (encoding the smallest details) and repeat the procedure, working this time with the signal seen at half resolution. This smoothed signal is divided into two more parts: an even smoother signal—seen at a quarter the resolution of the original—and new details, twice as big as the first. To do this we double the size of the scaling function and the wavelet. This time the work goes twice as fast; we have half as many coefficients to compute, and each coefficient is no harder to compute. The new details encoded by the wavelets and the average values computed with the scaling function are based on segments of the signal that are twice as big, but we're working now with the smoothed signal; the bigger segments contain the same number of points as the smaller segments of the first step.

The third step again takes half as much time as the second. We compute half again as many coefficients to produce a signal at one eighth the resolution of the original. Eventually, we will have smoothed the signal out of existence: all the information will have been drained off into wavelet coefficients, which are classified by resolution, each resolution corresponding to a certain scale, certain frequencies. If we stop before the signal has completely evaporated, the remaining information is encoded by the scaling function. (How this algorithm works for two-dimensional signals—pictures, for example—is discussed in *Wavelets in Two Dimensions*, p. 193.)

Beyond Plain English 9

Wavelets in Two Dimensions

When wavelets compress pictures, a fast approach is to use "separable products" of a one-dimensional wavelet and a one-dimensional scaling function. This makes it possible to use the fast wavelet transform, but favors the horizontal and vertical directions, which can be a drawback in analysis. Several different approaches make it possible to orient two-dimensional wavelets in any direction: a continuous two-dimensional transform, a discrete, oversampled transform that uses "steerable" filters, and special waveforms called *brushlets*, which combine steerability and orthogonality.

See p. 193.

The fast algorithm that we have just sketched—the FWT, or fast wavelet transform—is very close to the pyramid algorithms of Burt and Adelson, using filter banks. "Rereading them now, one wants to say, everything was there," says Meyer.

Now that we see it from a certain distance, I would say that an epsilon has been added to the magnificent work of Burt and Adelson. But the epsilon was still important. There was no scientific theory explaining the choices they made instinctively; something was missing on a scientific level.

They were lacking . . . the concept of the wavelet . . . and they didn't have the orthogonal aspect. They calculated coefficients, but the coefficients weren't interpreted as an orthogonal basis. But they have a tremendous instinct. . . . With hindsight a lot of justifications have been found for the choices Burt and Adelson made.

Like the FFT, the FWT is not a luxury that makes computations a little faster. It makes computations possible that otherwise would not be done at all. "For most one-dimensional signals, speed isn't essential; today you can compute very fast with specialized processors. But this becomes completely false when you talk about pictures in two dimensions, not to mention higher dimensions; then there are things that you just can't do without fast algorithms," Grossmann says.

Even in one dimension, Mallat says, speed can be indispensable: "If, for example, you want to calculate the wavelet transform of a speech signal in real time, you have to be able to compute 16 thousand samples a second . . . you absolutely have to have fast algorithms."

Beyond Plain English 10

Pyramid Algorithms of Burt and Adelson

"I suspect that no one will ever use this algorithm again," wrote one reviewer about the pyramid algorithms that inspired the fast wavelet transform.

See p. 199.

Evading Infinity: Daubechies Wavelets

Mallat first proposed his fast algorithm using truncated versions of infinite wavelets constructed by Guy Battle and Pierre Gilles Lemarié (now Lemarié-Rieusset). A new kind of orthogonal wavelet with *compact support* made it possible to avoid the errors caused by this truncation. These wavelets, constructed by Daubechies, are not infinite; they have the value zero everywhere outside a certain interval or *support* (between –2 and 2, for example). They are also, unlike Morlet's or Meyer's wavelets, creatures of the computer age. Daubechies's wavelets cannot be constructed from analytic formulas; they are made using iterations.

Beyond Plain English 11

Multiwavelets

A Daubechies wavelet is the limit of an iterative process and cannot be created from analytic formulas. Other researchers have found that one *can* make orthogonal wavelets with compact support that are explicit functions, by using more than one scaling function. The resulting multiresolution analysis is not subject to the limitations of Daubechies wavelets; it is possible, for example, to create symmetrical orthogonal wavelets with compact support. Two procedures for creating such multiwavelets are used, one producing "fractal" wavelets, the other piecewise polynomial wavelets.

See p. 201.

An iterative process consists of repeating the same operation, using each time the last output as the new input: iterating $x \mapsto x^2 - 1$ starting with $x = 2$ gives first 3, then 8, then 63 ... The fast wavelet transform uses iterations, since at each stage one uses the most recently smoothed version of the signal as the new starting point. Iterating is what a computer does best: "You give a single command and then you loop; it goes very fast," says Meyer.

Such iterations look deceptively simple. Nonlinear iterations like $x \mapsto x^2 - 1$ are the equivalent, in discrete time, of nonlinear differential equations—so difficult that for a long time mathematicians, finding prudence the better part of valor, avoided thinking about them. But over the past dozen years or so, they have learned to use computers to study iterative processes, creating extraordinarily complex objects like the Julia set and the Mandelbrot set from very simple computer programs, and producing, in the process, the beautiful pictures of "chaos" and dynamical systems.

Nonetheless, many mathematicians are still uneasy with iterations. In particular, says Meyer, it doesn't occur to them to use iterations to construct explicit functions. "On the other hand," he says, "for people who work with computers, who do signal processing, iterative methods are extremely natural." So Mallat, who used computers in his studies on artificial vision, had the idea of using iterations to construct wavelets (an idea also inspired by the pyramid algorithms of Burt and Adelson), and in his article on multiresolution, he had suggested this approach. But he didn't carry out the idea to its conclusion. "Mallat launches brilliant ideas that keep two or three hundred people busy, then he goes on to something else," says Meyer. "It was Ingrid Daubechies, with her tenacity, her capacity for work—and for subtle and difficult analysis—who implemented it."

Mallat recalls his decision not to attempt to create wavelets from his filters:

> After realizing that the orthogonal wavelet transform could be computed with a fast algorithm, using "special" digital filters, I thought that I had discovered a new and particularly interesting class of filters. At that time, I was enrolled in an electrical engineering Ph.D. program, although I was doing my thesis in computer vision. So I went to announce the news to Rashid Ansari, with whom I was doing an independent study on multirate filter banks. He pointed me to the work of Smith and Barnwell, who had come up earlier with the same conjugate mirror filter condition, which was very disappointing to me. I also realized that they had design techniques for building compactly supported conjugate mirror filters. Several other people in signal processing had been working on this topic, and so I thought that studying the design of such filters was not of great interest.
>
> Daubechies thought differently, and she was right because the filters of Smith and Barnwell had no vanishing moments and thus did not generate finite energy wavelets.[1]

---

[1]The connection between the number of vanishing moments of a filter and the corresponding wavelet having finite energy is not immediately apparent. But having a conjugate mirror filter h such that its Fourier transform $\hat{h}$ vanishes (is zero) at $\omega = \pi$ is a necessary condition so that the cascade of such filters defines a finite energy scaling function and hence a finite energy wavelet. In addition, the number of vanishing moments of a wavelet

Daubechies, who is Belgian, was trained as a mathematical physicist; she had worked with Grossmann in France on her Ph.D. research, then worked in the United States on quantum mechanics. She is the recipient of a five-year MacArthur Fellowship. At that time she was at the Courant Institute, after which she worked for AT&T. "Ingrid's role has been crucial," Grossmann says. "Not only has she made very important contributions, but she has made them in a form that was legible, and useful, to various communities. She is able to speak to engineers, to mathematicians, she is trained as a physicist and one sees her training in quantum mechanics." Daubechies had heard about Meyer's and Mallat's multiresolution work very early on. "Yves Meyer had told me about it at a meeting. I had been thinking about some of these issues, and I got very interested," she said. With Meyer's infinite orthogonal wavelets, calculating a single wavelet coefficient was a lot of work; Daubechies wanted to construct wavelets that would be easier to use. She was demanding. In addition to orthogonality and compact support, she wanted very smooth wavelets with vanishing moments: constraints so much in conflict that many experts doubted such wavelets were possible. (For some examples of wavelets, see Figure 2, p. 51.)

> I said why can't we just start from the fact that we want those numbers and that we want a scheme that has these properties, proceed from there. That's what I did.... I was extremely excited; it was a very intense period. I didn't know Yves Meyer so very well at the time. When I had the first construction, he had gotten very excited, and somebody told me he had given a seminar on it. I knew he was a very strong mathematician and I thought, oh my God, he's probably figuring out things much faster than I can.... Now I know that even if that had been true, he wouldn't have taken credit for it, but it put a very strong urgency on it; I was working very hard. By the end of March 1987 I had all the results.

---

is equal to the number of zeros of the Fourier transform of its filter at $\omega = \pi$: saying that a wavelet has one vanishing moment is equivalent to saying that $\hat{h}(\pi) = 0$. (More generally, if a wavelet has p vanishing moments, then $\hat{h}$ and its first $(p-1)$ derivatives vanish at $\omega = \pi$.)

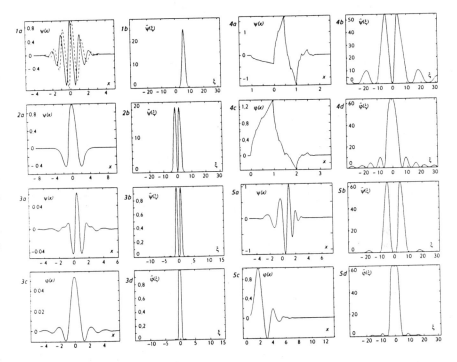

Figure 2. Examples of wavelets and the corresponding scaling functions and Fourier transforms. 1(a) The Morlet wavelet and 1(b) its Fourier transform. The Morlet wavelet is a complex function; the imaginary part is shown by the dotted line. 2(a) The "Mexican hat" wavelet and 2(b) its Fouriet transform. Both these wavelets are used in the continuous wavelet transform, which requires no scaling function. The other wavelets shown generate orthogonal bases. 3(a) The Meyer–Lemarié wavelet of order 4 and 3(b) its Fourier transform, 3(c) its scaling function, and 3(d) the Fourier transform of its scaling function. 4(a) A Daubechies wavelet, order 2; 5(a) a Daubechies wavelet, order 7. The order measures regularity: 5(a) is smoother than 4(a). The corresponding scaling functions are shown in 4(c) and 5(c); the Fourier transforms, in 4(b) and (d) and 5(b) and (d). (Courtesy of Marie Farge and Eric Goirand.)

Many applications of wavelets depend in a crucial way on their ability to encode signals with just a few nonzero coefficients. How well a wavelet does this depends in part on the signal to be encoded, but it also depends on two characteristics of the wavelet: the length of its support and the number of its vanishing moments. In addition, a small support implies a small number of operations in the fast wavelet transform: the number of operations is approximately $2kn$ for a signal of size n where k is the support size of the wavelet. Among orthogonal wavelets, Daubechies' wavelets provide the smallest possible support for a given number of vanishing moments, so they were of great importance for applications. "... it was the paper by Ingrid Daubechies in 1988 that caught the attention of the larger applied mathematics communities in signal processing, statistics, and numerical analysis," write C. Sidney Burrus, Ramesh Gopinath, and Haitao Guo in *Introduction to Wavelets and Wavelet Transforms: A Primer* [Burris et al.].

Heisenberg

Multiresolution and Daubechies's wavelets with compact support meant more than speed. They made it possible to analyze a signal in both time and frequency with unprecedented ease and accuracy, zooming in on very brief intervals of a signal without losing too much information about frequency. But although one mathematician hearing Daubechies lecture objected that she seemed "to be beating the uncertainty principle," the Heisenberg uncertainty principle still holds. Just as an elementary particle *does not simultaneously have* a precise position and a precise momentum, a signal does not simultaneously have a precise "location" in time and a precise frequency.

A very brief signal, well localized in time, necessarily has a Fourier transform that is spread out: a broad range of frequencies. Think of a very brief signal—zero everywhere except at a skinny peak. To represent it as a Fourier series, one would need to carefully add together a great many sines and cosines. Conversely, a signal with a very narrow range of frequencies is necessarily spread out in time; it's not possible to convince just a few sines and cosines to cancel out so that the value of the signal is zero outside of a narrow time interval.

As Heisenberg showed, the product of the two "uncertainties," or spreads of possible values—$\Delta t \, \Delta f$ ("delta t, delta f")—is always at least a certain minimum number. (This product does not depend on physical measurements, since frequency is the inverse of time; it is usually measured in cycles or in radians, and the exact number depends on this choice, purely a matter of convention. Physicists, on the other hand, multiply the uncertainty in position by the uncertainty in momentum, which is not the inverse of position; physical measurements enter into the formula and physicists come up with a very different number.)

Beyond Plain English 12

The Heisenberg Uncertainty Principle and Time-Frequency Decompositions

Most people first learn about the Heisenberg uncertainty principle in connection with quantum mechanics, but it is also a central statement of information processing. Any method of encoding information that relies on the fact that information can be broken down by frequency comes up against this limitation, which says that precision about time imposes a certain vagueness about frequency, and vice versa. The classical Fourier transform forces one to choose either time or frequency; windowed Fourier analysis and wavelets offer different compromises.
This limitation does not result from some mysterious shielding of reality from our eyes; it is reality, as we see by looking at the mathematics of Fourier analysis.

See p. 203.

Gabor was one of the first to insist on the relevance of the uncertainty principle to communication theory and acoustics. In his 1946 paper he suggested that although popular articles about quantum theory had made the Heisenberg principle widely known, the actual mathematical statement setting a lower limit to the product $\Delta t \, \Delta f$ "has received less attention than it deserves" [Gabor, p. 432]. In communication theory, he added, its importance "appears to have passed unnoticed."

Wavelets can't overcome this limitation, although they adapt automatically to a signal's components—becoming wider to analyze low frequencies, skinnier to analyze high frequencies. "At low frequencies I have wide wavelets, and I localize very well in frequency but very badly in time," says Daubechies. A big wavelet, necessary to accommodate low frequencies, is forced to be vague about time; it produces a single coefficient that corresponds to a relatively long time span.

At very high frequencies the problem is the reverse: "I have very narrow wavelets, and I localize very well in time but not so well in frequency," Daubechies says. No matter how hard you squeeze the wavelets to encode smaller and smaller intervals of the signal, the frequencies slip out of the embrace; each time you go to a scale twice as small, you go up an octave and the range of frequencies doubles.

This widening spread of frequencies can be seen as a barrier to precision, but it's also an opportunity that engineers have learned to exploit. It's the reason the telephone company shifts voices up into higher frequencies, not down to lower frequencies, when it wants to fit a lot of voices on one line: there's a lot more room up there. It also explains why fiber optics, which carry high-frequency light signals, are more efficient than conventional telephone wires.

Heisenberg, then, imposes a compromise; knowledge gained about time is paid for in frequency, and vice versa. Even talking about wanting to localize a signal in time and frequency can be misleading, since it seems to suggest that each signal has a single correct time-frequency decomposition—precise frequencies for each instant—that is hidden from us. But how can one talk about frequencies at a precise instant, when frequencies have to have the time to oscillate?

A signal's time-frequency decomposition isn't given once and for all; it depends on the point of view of the observer. Interrogate the signal in one way (with windowed Fourier, for example), and it will give one answer, always with the ambiguity imposed by the Heisenberg principle. Interrogate it in another way (with wavelets, perhaps), and it will give a different answer, just as correct, maybe more useful, maybe less. Although this multiplicity of perspectives may seem strange at first, it is not fundamentally more mysterious than the fact that the number 24 can be factored in different ways. But in quantum mechanics, this same mathematical statement produces staggering consequences.

Beyond Plain English 13

Probability, Heisenberg, and Quantum Mechanics

The Heisenberg uncertainty principle is often interpreted to mean that we can't simultaneously know the position and momentum of an elementary particle. The truth is stranger: an elementary particle does not simultaneously have a precise position and a precise momentum. Fourier analysis is the language we need to express this peculiar situation. Mathematically, position and momentum correspond to the two different sides of the Fourier transform.

Quantum mechanics is probabilistic, but in most cases instead of dealing in discrete probabilities, like the odds of getting a straight flush or getting heads 10 times in a row, these probabilities are continuous. We can calculate the probability that a particle will be in a given region, but not the probability that it will be at a particular point. Integrals are the natural tools to use to express these continuous probabilities.

See p. 209.

In quantum mechanics, answers also depend on the way the question is asked; i.e., on the experiment performed. When physicists asked light what it was made of, sometimes it answered that it was made of particles, sometimes that it was made of waves. But while time-frequency decompositions are different perspectives on a signal that one knows, and that doesn't change, one can't have a perspective on the reality of quantum mechanics without changing that reality. Every time one measures the system, one changes it. So saying that light is made of particles or of waves makes no sense. It's both at once, in flagrant contradiction with our experience and common sense. There is only one reality, the wave function; trying to understand it with a mentality molded by our deterministic world seems doomed to fail.

CHAPTER IV

Applications

Wavelets appear highly unlikely to have the revolutionary impact upon pure mathematics that Fourier analysis has had. "With wavelets it is possible to write much simpler proofs of some theorems," Daubechies says. "But I know of only a couple of theorems that have been proved with wavelets, that had not been proved before." (Strichartz adds that this simplification is in itself not negligible.) But wavelets are suited to a wide range of applications. It can be instructive to compare wavelet coefficients at different resolutions. Zero coefficients, which indicate no change, can be ignored, but nonzero coefficients indicate that something is going on, whether an error, an abrupt change in the signal, or noise (an unwanted signal that obscures the real message). If coefficients appear only at fine scales, they generally indicate the slight but rapid variations characteristic of noise. "The very fine-scale wavelets will try to follow the noise," Daubechies explains, while wavelets at coarser resolutions are too approximate to pick up such slight variations.

But coefficients that appear at the same part of the signal at all scales indicate something real. If the coefficients at different scales are the same size, this indicates a jump in the signal; if they decrease, it indicates a singularity—an abrupt, fleeting change. It is even possible to use scaling to sharpen a blurred signal. If the coefficients at coarse and medium scales suggest there is a singularity, but at high frequencies noise overwhelms the signal, one can project the singularity into high frequencies by restoring the missing coefficients—and end up with something better than the original. Indeed, image enhancement by selectively changing certain coefficients before reconstructing an image has been used to improve mammograms; since the contrast between the soft tissues of the breast is

57

small, it can be easy to miss a small change indicating a malignant tumor [Unser, Aldroubi, p. 632].

Beyond Plain English 14

Traveling from One Function Space to Another:
Wavelets and Pure Mathematics

Fourier's realization that even a discontinuous curve can be represented as a sum of sines and cosines and thus treated like a function contributed to a profound and wrenching change in mathematics. Over the course of the nineteenth century, mathematicians discovered strange new functions, such as the function

$$f(x) = \cos 2\pi 10x + \frac{1}{2}\cos 2\pi 10^2 x + \frac{1}{3}\cos 2\pi 10^3 x + \frac{1}{4}\cos 2\pi 10^4 x + \cdots,$$

which jumps constantly from minus infinity to infinity, but gives a finite value for any number x chosen at random. They also developed new tools, such as the Lebesgue integral, to deal with these strange creatures.

The enlargement of the notion of function was followed by another profound change. Instead of thinking of individual functions, mathematicians began studying whole families of functions (and even odder creatures called distributions) and talking of the "function spaces" they inhabit. Because wavelet coefficients, unlike Fourier coefficients, faithfully reflect the properties of functions (at least, Yves Meyer qualifies, "everything that is rupture, discontinuity, the unforeseen"), they are a useful tool in studying these function spaces. With a simple change of coefficients, one can travel from the smooth slopes of ordinary functions to the peaks and chasms of "the wildest distributions."

See p. 223.

Because wavelets respond to change and can narrow in on particular parts of a signal, researchers at the Institut du Globe in Paris are using them to study the minuscule effect on the speed of the earth's rotation of the El Niño ocean current, which flows along the coast of Peru, while researchers at the University of Southampton are using wavelets to study ocean currents around the Antarctic [Sinha, Richards]. In mechanics, researchers are exploring the use of wavelets to detect faults in gears, by analyzing vibrations.

The fact that errors in a wavelet transform don't corrupt the entire transform can be useful in medical imaging. When Fourier analysis is used in magnetic resonance imaging, any movement of the organ being studied distorts the whole image. Mathematician Dennis Healy, Jr. and radiologist John Weaver at Dartmouth College have found that with wavelets, these "motion artifacts are dramatically reduced" [Healy, Weaver, p. 849]. They are also exploring using multiresolution in "adaptive" magnetic resonance imaging, in which one would use high frequencies selectively, depending on results obtained at lower frequencies. (This could led to important savings; half an hour of MRI can cost more than $1000.)

Multiresolution is also a promising approach for comparing functional magnetic resonance images of the brain, acquired under different conditions: instead of guessing at the best single resolution, researchers can use wavelets to analyze the signal at many resolutions [Ruttimann et al.]. Researchers at the National Institutes of Health have also explored the use of wavelets in positron emission tomography (PET), for example in studying differences in brain function of alcoholic patients and healthy volunteers [Unser et al. 1].

And researchers at the University of Rochester Medical Center have found wavelets to be a valuable tool in high resolution electrocardiography, with potential both for predicting patients at risk of sudden death following myocardial infarction, and for detecting abnormalities characteristic of "long QT syndrome," a rare disorder thought to cause 3,000 to 4,000 deaths a year in the United States among children and young adults [Couderc et al.].

In astronomy, wavelets are being used to study the large-scale distribution of matter in the universe, which for years was thought to be random but which is now seen to have a complicated structure, with "voids" and "bubbles." A better understanding of this structure is necessary, writes Albert Bijaoui, in order to evaluate various scenarios for the evolution of the universe; the first step is to make an inventory of the universe, counting galaxies [Bijaoui, p. 195].

Since wavelets make it possible to identify structures at different scales, they can distinguish between a star and a galaxy, which isn't always obvious. They enabled Bijaoui and his colleagues at the Observatory of the Côte d'Azur in Nice to identify a subcluster at the center of the Coma supercluster, a cluster of about 1400 galaxies. That subcluster was subsequently identified as an x-ray source. "Wavelets were like a telescope pointing to the right place," Meyer says.

The problem of analyzing the shapes of neural cells is another field where it is hoped that wavelets will be a useful tool. Biological neural systems involve from 50 to 500 distinct types of neurons. "Their shape is particularly important since it determines a great deal of their functional properties," says Luciano da Fontoura Costa of the Cybernetic Vision Research Group at the University of Sao Paulo in Brazil, who is working on the problem with Roberto Marcondes Cesar, Jr. [Cesar, Costa], [Cesar et al.]. In addition, better analysis of neuron shape would be useful in modeling such cells and in automatic classification of neurons based on shape.

The task is a difficult one. There is little agreement among researchers on how to measure neuron shape, or what aspects are most important; in addition, Costa says, neurons are generally irregular in shape, images of neurons are generally contaminated by noise, and attempts to analyze the shape by analyzing curvature run into mathematical difficulties. In estimates of the curvature of circular contours with standard techniques, errors ranging from 1% to 1000% have been reported [Worring, Smeulders].

He and Cesar are hopeful that a multiscale approach with a continuous wavelet transform will provide a unified approach to deal with the different problems in shape analysis—particularly since wavelet theory has a sound mathematical background [Antoine et al.]. If so, Costa says, it will be a natural development. "One of the most classical approaches for shape representation is through the so-called *Fourier descriptors*, which can be obtained from the Fourier transform of a signal that represents the shape. But Fourier descriptors suffer from the same general problems of Fourier analysis, namely its global nature and inability to deal with transient features of signals. So the wavelet approach for shape analysis can be thought of as the natural evolution of the well-established Fourier descriptors."

The Construction Rule of Fractals

Wavelets seem particularly well suited to the job of understanding fractals, or multifractals, often characterized by self-similarity at different scales.[1] Actually, the question of knowing whether wavelets are the right tool for studying these objects illustrates how fast the situation can change.

"A true mathematical microscope, this transformation invites us to descend into the hierarchical edifice of fractals," wrote Alain Arnéodo, F. Argoul, and G. Grasseau in 1990, inviting their readers to a "voyage into the heart of fractals" and talking of "stripping bare the construction law of critical fractals...." [Arnéodo et al., p. 127]. Two years later, Meyer suggested that another method for studying fractals was more flexible and precise [Meyer 3, p. 118]. But in 1993, results obtained with the "turbulence signal" made him change his mind. "The situation is extremely satisfying for the person who is living the adventure—and extremely humiliating as soon as one writes a book," he remarks.

The signal of turbulence is a signal that appears completely chaotic, kilometers of which have been measured in the wind tunnel at Modane, in France near Italy. A "random number generator" can give a series of apparently random numbers; with simple computer programs, researchers in dynamical systems have created very complex sets. Can one do the reverse and find a simple rule for a complex signal? Does the signal of turbulence have a hidden structure? Very likely it does, as it is determined by the relatively simple Navier-Stokes equations; but finding that structure is a formidable task. A method using wavelets, and tested by Arnéodo

---

[1]While the word "fractal" applies to sets, the word "multifractal" applies only to measures and functions (except in some very special cases). In general, explains Uriel Frisch, a multifractal function is characterized by various types of singularities on various fractal sets: for example, discontinuities of the function on a set A of dimension a and discontinuities of the derivative of the function on a set B of dimension b. This is an example of a "bifractal" function; for a multifractal function, one should imagine an infinity of such singularities and corresponding sets.

and his colleagues at Bordeaux, established that the Modane signal has a multifractal structure; Stéphane Jaffard, at the French Ecole Nationale des Ponts et Chaussées, has found a mathematical justification for it. "Suddenly we're back to wavelets as the method best suited to the job at hand," says Meyer.

But Uriel Frisch of the Observatory of Nice, who along with Giorgio Parisi of the University of Rome La Sapienza created the multifractal model used in this work [Parisi, Frisch], says he is not "one hundred percent convinced that the model is pertinent" for signals like the one at Modane. The signal, though turbulent, may not be turbulent enough. "The instances where observations diverge from the self-similarity assumed by Kolmogorov's 1941 theory ... could result from an insufficient Reynolds number," he remarks. (The Reynolds number of a signal is a measure of its turbulence.) It is possible that Kolmogorov's theory is correct in the limit of an infinite Reynolds number, but not for Reynolds numbers on the order of a million, like that characterizing the Modane experiments. The differences observed could also be artifacts, he says; a Brownian motion with ultraviolet cutoff, which is continuous, "can camouflage itself in places as an almost discontinuous process," a phenomenon he calls the *chameleon effect*. "If you try to extract local exponents of a process using a wavelet analysis that is too naive, you will see all sorts of artifacts for this reason." A new method for analyzing turbulence data, "Extended Self Similarity," supports the multifractal hypothesis, he says, "but even so there is still room for doubt."[2]

Cut the Weeds and Spare the Daisies: Wavelets for Denoising

Wavelets also made possible a new method for extricating signals from white noise ("all-color," or all-frequency, noise), which has great potential in many fields, including medical scanning and molecular spectroscopy.

---

[2]Extended Self Similarity was developed by Roberto Benzi in Rome and S. Ciliberto at the Ecole Normale Supérieure in Lyon, and by their co-workers. See [Benzi et al.].

An obvious problem in separating noise from a signal is knowing which is which. (Even defining noise isn't always obvious; to someone searching for undersea oil, a submarine is noise, points out Morlet, while to the military, the same submarine is the signal.) If you know that a signal is smooth—changing slowly—and that the noise is fluctuating rapidly, then you can filter out noise by averaging adjacent data to kill fluctuations while preserving the trend. This filtering won't blur smooth signals too much, since they are mostly lower frequency.

But many interesting signals (the results of medical tests, for example) are not smooth; they contain high-frequency peaks. Killing all high frequencies mutilates the message—"cutting the daisies along with the weeds," in the words of Victor Wickerhauser of Washington University in St. Louis.

A simple way to avoid this blind slaughter has been found by a group of statisticians. David Donoho of Stanford University and of the University of California at Berkeley and Iain Johnstone of Stanford had proved mathematically that if a certain kind of orthogonal basis existed, then it would do the best possible job of extracting a signal from white noise. (A *basis* is a family of functions with which you can represent any possible function in a given space; each mother wavelet provides a different basis, for example, since any function can be represented by it and its translates and dilates.)

This result was academic, since Donoho and Johnstone did not know whether such a basis existed. But in the summer of 1990, when Donoho was teaching a course in probability in St. Flour, in France's Massif Central, he heard Dominique Picard of the University of Paris-Jussieu give a talk on using wavelets in statistics. After discussing it with her and with Gérard Kerkyacharian of the University of Picardy in Amiens, "I realized it was what we had been searching for for a long time," Donoho recalled. "We knew that if we used wavelets right, they couldn't be beaten."

The method is simplicity itself. You apply the wavelet transform to your signal, throw out all coefficients below a certain size—at all resolutions—and then reconstruct the signal. It is fast, because the wavelet transform is fast, and it works for a variety of kinds of signals (see Figure 1; also [Donoho, Johnstone] and [Donoho et al.]).

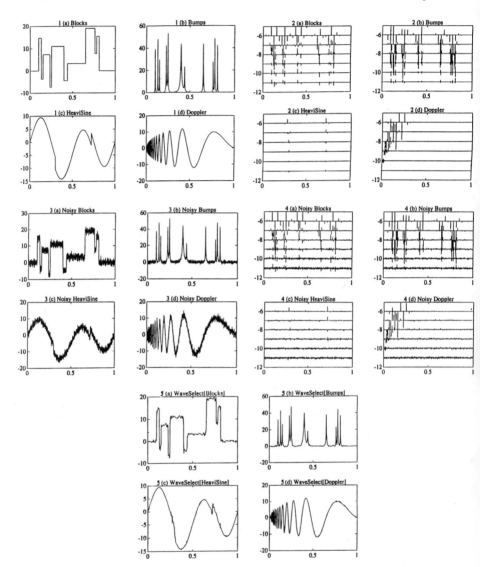

Figure 1. Wavelet Thresholding to Tackle White Noise. Wavelets can be used to remove white noise that is contaminating signals: all wavelet coefficients below a certain threshold are eliminated. 1. Four functions   2. The same functions in wavelet space (i.e. their wavelet coefficients)   3. The functions with white noise 4. The functions with white noise, in wavelet space. A few large coefficients stand out against a background of noise   5. The functions reconstructed from the largest coefficients. (Courtesy of David Donoho and Iain Johnstone.)

The surprising thing is how little one needs to know about the signal. The traditional view is that one has to know, or assume, something about the signal one wants to extract from noise—that, as Grossmann put it, "if there is absolutely no *a priori* assumption you can make about your signal you may as well go to sleep. On the other hand you don't want to put your wishes into your algorithm and then be surprised that your wishes come out." In particular, traditional methods require knowing or assuming something about how smooth a signal is—whether it belongs to the class of signals that are smooth except for sudden jumps, for example, or to the class of signals with very sharp peaks that decrease rapidly, or perhaps to the class of chirps (with a sliding frequency, like an admiring whistle).

With wavelets, it's enough to know that a signal belongs to a much larger family, which includes these classes and many more—essentially all signals to which standard denoising methods are applied. Then, without knowing anything more, Donoho says, "you do as well as someone who makes correct assumptions, and much better than someone who makes wrong assumptions."

The trick is that for this large family of signals, an orthogonal wavelet transform compresses the "energy" of the signal in a relatively small number of big coefficients; it classifies the signal into a few compartments. But it can't "classify" white noise. As Meyer says, "white noise is completely disordered; it's feverish and it will remain feverish no matter what system of representation you use." (That all orthogonal representations leave white noise unchanged has been known since the 1930s.) The energy of white noise is dispersed throughout the transform, giving relatively small coefficients, which are thrown out. So while noise masks the signal in "position space," the two become disentangled in "wavelet space."

Other researchers, including Ronald DeVore at the University of South Carolina and Bradley Lucier at Purdue University [DeVore, Lucier] and Mallat at Courant, independently discovered that thresholding wavelet coefficients is a good way to kill white noise. "We came to it by mathematical decision theory, others by approximation theory, or simply by working with wavelet transforms and noticing what happened," Donoho remarks.[3]

Mallat was interested in blurred images. Some statisticians suggest that while Donoho's and Johnstone's method is elegant and important, it may be more impressive in theory than in practice: a method adapted to a particular problem may be preferred over one that is "near optimal" for a whole range of problems (see, for example, the discussion following [Donoho et al.]). When it is applied to blurred images, it damages some of the edges, creating annoying ripples. Mallat and graduate student Wen Liang Hwang [Mallat, Hwang] developed a way to avoid this, shown in Figures 2 and 3[4]. After computing a signal's wavelet transform, they choose those coefficients that are bigger than neighboring coefficients—corresponding, at each scale, to points of the image where the correlation between the image and the wavelet is greatest, compared to neighboring points. Since wavelets react to variations—in images, to edges—these maximal values, or *wavelet maxima*, correspond in principle to the points that make up edges. Those wavelet maxima that seem to correspond rather to noise are discarded; that decision, based on the existence and size of maxima at different resolutions, is made automatically but requires more computations than Donoho's method.

---

[3]Edward Adelson relates that he described this method of wavelet denoising in 1986, in a report he wrote while working at RCA labs; the company declined to patent it, but would not let him publish it. The unpublished paper is now available at his web site [Adelson]; see also [Simoncelli, Adelson]. "I am sorry that I couldn't publish it early on, but it might have received little attention if I had," he says. "For applications people, the computational costs were a problem in 1986. And the theorists weren't interested because there was no elegant math." The history of the technique goes back to work on two-band coring by TV engineers in the 1950s; by the 1980s, engineers had extended it to 2-D coring, orthogonal bases, oriented coring, and multiscale coring. Adelson's own work was an attempt, he says, to redo in a cleaner manner work by two Kodak engineers in the 80s. "The Kodak methods were real breakthroughs, working far better than classical coring or the classical academic techniques based on statistical theory. However, the processing requirements were too great for practical use at that time, and people in academia were largely unaware or uninterested. Today, there is satisfying math to excite the theorists, which spawns activity and interest, which in turn excites the applications engineers."

[4]The woman in Figure 3 (p. 68) is often identified in the image processing literature as Lena. Her name is actually Lenna Sjöblom; the picture appeared in the November 1972 issue of *Playboy* magazine. Recently, Ms. Sjöblom was a guest of honor at a symposium on digital signal processing [Frazier, Chapter 3].

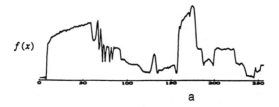

$f(x)$

a

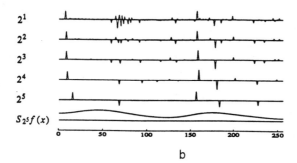

2^1

2^2

2^3

2^4

2^5

$S_{2^5}f(x)$

b

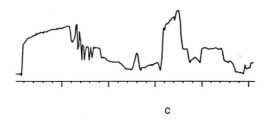

c

Figure 2. Wavelet Maxima. A signal can be represented by its wavelet maxima—those wavelet coefficients that are large compared to nearby wavelet coefficients. (a) The original signal (whose wavelet transform is shown in Figure 3, Chapter II, p. 33). (b) The wavelet maxima representation of the signal. (c) The signal reconstructed from the wavelet maxima, using 20 iterations. (Courtesy of Stéphane Mallat.)

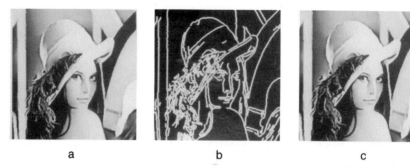

a   b   c

Figure 3. "Plastic surgery" with Wavelet Maxima. Wavelets were used to compute the edges of the image (a) of Lenna, at different resolutions; wavelet maxima then chose the most "important" ones. The edges at a fine resolution are shown in (b). In (c), the image was reconstructed from the edges at various scales, using an algorithm developed by Stéphane Mallat and Sifen Zhong. The edge selection gives Lenna a face lift: it removes fine details, making the skin much smoother. (Courtesy of Stéphane Mallat.)

Wavelet maxima were developed not just for denoising, but as an automatic way to detect and characterize singularities—places in a signal "where something interesting is happening," says Mallat—whether for denoising, encoding, or analysis. They are used, for example, by Arnéodo and his colleagues to analyze singular structures in the velocity field of a turbulent fluid, as well as to analyze certain fractals [Arnéodo et al. 2], and to study the fractal scaling organization of DNA sequences [Bacry et al.]. They have also been used in magnetic resonance images, where they have produced sharper edges without inducing ringing artifacts (as Gibbs phenomena are known in image processing) [Xu et al.]. One advantage, important in recognizing patterns, is that the wavelet maxima representation is translation invariant; where on the signal one starts the encoding doesn't matter.

Shooting Yourself in the Foot: Artifacts and Other Hazards

Although Donoho's and Johnstone's technique is simple and automatic, wavelets aren't foolproof. With orthogonal wavelets it matters where one starts encoding the signal: shifting over a little changes the coefficients completely, making pattern analysis hazardous. This danger does not exist with a continuous transform, but such a transform holds its own pitfalls.

What looks like correlation of coefficients (different coefficients "seeing" the same part of the signal) may be an artifact introduced by the wavelets. "It's the kind of thing where you can shoot yourself in the foot without half trying," Grossmann says.

Using wavelets requires practice. "With a Fourier transform, you know what you get," Meyer says. "With a wavelet transform you need some training in order to know what you get. I have a report from EDF [Electricité de France] giving conclusions of engineers about wavelets— they say they have trouble with interpretation." Part of that difficulty may be fear of trying something new; Gregory Beylkin of the University of Colorado at Boulder reports that one student, who learned to use wavelets before knowing anything about Fourier analysis, experienced no difficulty; Farge has had similar experiences. But Meyer thinks the problem is real.

Because Fourier analysis has existed for so long, and because most physicists and engineers have had years of training with Fourier transforms, interpreting Fourier coefficients is second nature to them. In addition, Meyer points out, Fourier transforms aren't just a mathematical abstraction; they have a physical meaning. "These things aren't just concepts, they are as physical, as real, as this table. But wavelets don't have a physical existence; that's why it is harder to interpret wavelet coefficients," he says.

"Physicists are sometimes reluctant to learn about wavelets because they cannot be interpreted in physical terms as easily as sines and cosines and their frequencies," writes J. C. van den Berg in the preface of *Wavelets in Physics* [van den Berg].

An orthogonal wavelet transform, which uses only wavelets dilated by a factor of two—thus an octave apart—is particularly arbitrary. That a signal (a piano sonata, for example!) can be represented using only the note C, or only the note B-flat, may seem ludicrous—just as some of Fourier's contemporaries may have viewed as absurd the idea that a discontinuous function can be represented as a superposition of sines and cosines. Yet the idea of decomposing information into components at different scales—into the main idea and details of different sizes—is in some sense very natural. And it appears in fact that our eyes and ears use a kind of wavelet analysis in the first stages of information processing.

When we see, neurons of the visual cortex are triggered by patterns of light called *receptive fields*. Just as narrow wavelets encode high frequencies and wide wavelets encode low frequencies, "neurons with small receptive fields respond to high frequencies and neurons with large receptive fields respond to low frequencies," says David Field, a psychologist at Cornell University. A similar phenomenon has long been known for hearing: *constant Q* filtering, in which the higher the frequency, the better resolution you have in time.

It is perhaps unwise to read too much into this connection. "I asked a friend who is a specialist in hearing how it works," Grossmann says. "He told me, if you had asked me three years ago I would have told you everything. But the more we know, the more we realize that things are really more complicated...." If, as is thought, our ears perform a multiresolution analysis in the first stages of processing sound, this multiresolution is followed by other, very complex processes. Adelson says he is

> disappointed that the wavelet revolution has had little impact on the field of vision research, which is my main interest. Gabor-like functions and multiscale representations such as pyramids are crucial in vision, of course, but we have known that for many years.
>
> I had hoped that the new mathematical tools would help us develop better models for human vision and better algorithms for machine vision. This has not occurred, in spite of some enthusiastic attempts; the new mathematics, which has received all the attention, has not enabled us to do anything that we couldn't already do with the old mathematics.

But the use of wavelet techniques by our eyes and ears may make wavelets particularly effective in compressing information. "If our ear uses a certain technique to analyze a signal, then if you use that same mathematical technique, you will be doing something like our ear," Daubechies says. "You might miss important things but you would miss things that our ear would miss too."

That opinion is widespread among wavelet researchers, says Field; he thinks it is wrong, at least for vision. Our visual system, he says, uses wavelets not for compression but to recognize what we see: to discriminate, for example, among hundreds of faces. "It's an optimal algorithm, but not for compression," he says.

Beyond Plain English 15

Wavelets and Vision: Another Perspective

Researchers in vision, working independently, created wavelets in the 1980s, resolving a debate between those who thought that cells of the visual system react to spatial frequency and those who thought they react to space. In a wavelet model of vision, the "image" we see is the signal, patterns of light (receptive fields) are the wavelets, and the reaction of a neuron to a particular pattern of light is the wavelet coefficient. As in a wavelet transform, small receptive fields give good information about space, while large receptive fields give good information about frequency.

One theory is that our visual system uses a "wavelet transform" to enable us to discriminate between different objects. When applied to natural scenes, only a small proportion of the available neurons fire at any given time. This means that a wavelet transform can encode natural scenes concisely, but David Field of Cornell University believes the goal is not compression but recognition.

See p. 231.

Quantifying Information

Why can't 10 hours of music be put on a single compact disc? Why aren't video telephones commonplace? The answer can be found in the sampling theorem. By showing that a continuous signal can be reproduced from a limited number of samples, this theorem made possible many technological marvels we take for granted. But it has a kicker: the amount of information a band of frequencies can carry in a given time is limited. Believing that a finite signal must be reproduced continuously is equivalent to believing that it contains an infinite amount of information. The sampling theorem shows that such signals contain far less information—and a telephone wire, for example, *transmits* far less information—than had been thought.

This realization "dawned gradually upon communication engineers during the third decade of this century," wrote Gabor; the sampling theorem confirmed and quantified their growing suspicion that ingenious attempts to increase the amount of information that could be transmitted in a given bandwidth involved "a fundamental fallacy" [Gabor, p. 429].[5]

One important consequence is the realization that information can be quantified. One can talk about the number of "bits" in the DNA that encodes a human being, for example. This has completely changed the way people think about information. (One "bit" refers to a choice between two alternatives: for a computer, 0/1, or a circuit that is off or on. The word was coined by John Tukey of the FFT from "b[inary] [dig]it." Eight "bits" make one "byte.")

Another consequence is the growing disproportion between the speed of computation and the speed of transmission. A computer (for example, a Power Mac) that can do some 20 million multiplications a second may be hooked up to a 9600 baud (bits per second) modem that can transmit 200 to 300 numbers a second. Computers relentlessly spew forth torrents of bits, flooding telephone lines and creating enormous bottlenecks; a bandwidth that is ample even for a multitude of conversations is pitifully inadequate to cope with the demands of computers.

(This discrepancy is just getting worse: computer speeds are increasing faster than modems can keep up; modems are already coming against the absolute limit imposed by bandwidth achievable on the telephone wiring used in most homes. Even with the much faster connection by Ethernet, transmission speed is a problem.)

One way to handle an ever-increasing volume of signals is to widen the electronic highways—for example, by moving to higher frequencies. Another solution, which also reduces storage and computational costs, is to compress the signal temporarily, restoring it to its original form

---

[5]In theory a single sample contains an infinite amount of information. But when measuring a signal, only a limited number of digits can be meaningful. Even with an ideal measuring device one would soon be looking at noise, not the signal—not to mention such problems, at the atomic level, as Brownian motion. The number π has been computed to a few billion digits, but for virtually all practical applications, $\pi = 3.14159$ is sufficient: 20 bits in base 2; the speed of light is one of the physical quantities that we know the most precisely, but it is known only to eight significant digits.

when needed. A simple approach to compression is to eliminate obvious redundancy or unimportant information, as people did to save money on telegrams, stripping a sentence to its essentials.

But one can also compress some signals that do not appear redundant, without losing any information. Signals can be protean, drastically changing their form (and often their apparent complexity) with the form of representation. An algebraic formula and its graph contain the same information; while the graph of a circle and its formula are both simple, consider graphing the function discussed in *Traveling from One Function Space to Another*, p. 223, which jumps constantly between plus and minus infinity. Storing or transmitting the formula is trivially easy; storing or transmitting the graph is impossible. In other cases, very simple instructions can encode very complicated output; the highly complex Mandelbrot set shown in Figure 4 can be produced by program about 30 lines long, and the information needed to create a human being is encoded by a relatively small amount of DNA (about a billion bits).

So the idea that information can be "quantified" requires amplification. How can we say how much information a signal contains if the same information appears to be very large or very small, depending on how a signal is represented? What determines whether a signal is compressible or not, that is, whether it has a more concise representation?

The approach of the Russian mathematician Andrei Kolmogorov was to define a signal's *information content* as the shortest sequence that encodes it in a given language (the computer language Pascal, for example). A signal that cannot be encoded by anything shorter than itself—that cannot be compressed—is by definition random.

This is at variance with the way the word *random* has traditionally been used in probability theory. In that view, a process can be random, but not the result of such a process. If you roll a 10-sided die 10 times, you will be just as unlikely to come up with 3,8,5,9,10,4,2,7,6,8 as with 2,2,2,2,2,2,2,2,2,2 or 1,2,3,4,5,6,7,8,9,10. But the first sequence would not surprise us, while the other two would: we treat the first sequence not as a unique and improbable sequence of numbers, but as representative of all such sequences with no apparent pattern. Faced with a signal, our first question is, does it mean something or doesn't it? In this we are

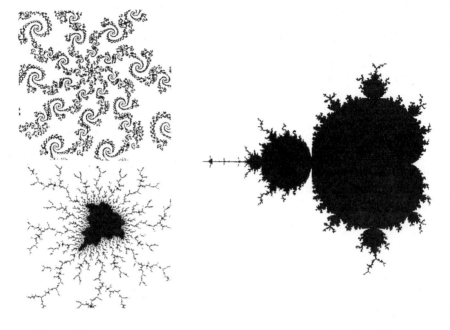

Figure 4. A program about 30 lines long can create the picture of the Mandelbrot set shown at right and myriad blow-ups of the set, two of which are shown at left. The program that generates these pictures can be stored with a few thousand bytes. Direct storage of the pixels would take about 32,000 bytes; compression techniques bring that figure down to fewer than 5,000 bytes for the picture at right, and between 14,000 and 25,000 for a typical blow-up. The number of different blow-ups generated by the same program is limited only by the necessity of specifying which blow-up you want.

The comparison between program and color pictures is more striking; the program that generates color Mandelbrot sets is much shorter, while the pictures contain more mathematical information.

The *Kolmogorov information content* of a signal is its shortest possible encoding in a given language (such as the programming language Pascal, which can encode the pixels of the picture or the instructions of the program). The enormous disparity between the Kolmogorov information content of some signals and their complexity in a different representation is of obvious interest. (Courtesy of Yuval Fisher.)

using a mathematical notion of meaning: does the signal have some kind of order or structure? Kolmogorov's use of the word "random" is thus very intuitive.

The use of the term "information content" is not. A random, incompressible signal has high information content, while a structured, compressible signal has low information content. Seen from one point of view this makes sense: if you want to send a signal over the telephone wires, a signal with low *Kolmogorov information content* will take up less room than one with high information content. But the terminology does violence to our feeling that information should mean something. The notion that the strings of letters *I love you* and *New York City has been destroyed by an atom bomb* contain less information than *z ugpw krb* or *ocy kwoh sitp kxu bvmw otldowycm le px ktow fxub*—because the first two can be compressed (I luv u; NYC ...) and the others cannot—seems unnatural. But "...*information* must not be confused with meaning," admonishes Warren Weaver [Shannon, Weaver, p. 8]. "...One has the vague feeling that information and meaning may prove to be something like a pair of canonically conjugate variables in quantum theory, they being subject to some joint restriction that condemns a person to the sacrifice of the one as he insists on having much of the other" [Shannon, Weaver, p. 28].

Naturally the most interesting instances of compression—where a very complicated signal is encoded in a very simple way—are those for which the shortest encoding is the hardest to find. You would be hard-pressed to deduce the program that creates the Mandelbrot set from the picture, let alone discover the structure of DNA by observing a human being.

In fact, as Kolmogorov pointed out in the 1960s, most signals are random and not compressible (see [Li, Vitanyi]). His argument is simple: using a given language, the number of short sequences is much smaller than the number of long sequences, so most long sequences cannot be encoded by anything shorter than themselves. Even a highly efficient encoding scheme like a library card catalogue cannot cope with all possible books of all possible lengths. Eventually the only way to distinguish one book from another would be to print the entire book in the card catalogue. (Even with a finite library, the larger the library, the more information one needs in order to locate a book; vaguely recalling the author's last name is no longer enough.)

In addition, even if a signal is compressible, Kolmogorov's message is discouraging. While statistical tests can uncover the structure of some compressible signals, for most signals there is no systematic way of even figuring out whether they are compressible or not. There is nothing smarter than trying all the possibilities.

Wavelets and Compression

Fortunately for people who work in signal processing, the signals that people want to compress generally have a structure that lends itself to compression. And rather than seeking the most succinct form for a signal, people tend to use general techniques that can be applied automatically to a whole class, or classes, of signals. As a general compression technique, wavelets have some built-in advantages. In many signals that people want to compress, a given point is more likely than not to be similar to points near it. (In a picture of a white house with a blue door, a blue point is likely to be surrounded by other blue points, and a white point by other white points.)

Wavelets are well suited to such signals. Because wavelet coefficients only indicate changes, areas with no change (or very small change) give small or zero coefficients, which can be ignored, reducing the number of coefficients that have to be kept to encode the information. But making effective use of wavelets in image compression required new approaches. In 1993, Jerome Shapiro, now with Aware, Inc., took advantage of the ability of wavelets to analyze a signal at different resolutions in his "embedded zerotree wavelet algorithm" for image compression [Shapiro]. An "embedded" code provides a rough approximation of the image, which is then progressively refined: when downloading a web site, one would like to quickly see what's there, even if the image is not sharp, rather than stare at the computer monitor while the picture slowly inches its way down from the top of the screen to the bottom.

It's possible to progressively transmit data using multiresolution analysis, sending coarse coefficients first, then finer and finer coefficients. But if the transmission is interrupted halfway through one particular resolution, the result will be a picture that is higher resolution at the top than at the bottom. With Shapiro's embedded code, the effect of an interrupted transmission is generally imperceptible: the picture you get after transmitting x bits is essentially the same picture that you would have gotten if you had planned to use only x bits all along. This is because bits are grouped according to their importance (mainly, their size), rather than by resolution. "Each time you add a bit, you are improving fidelity, not resolution," Shapiro says. "It's so gradual, it's hard to tell the difference."

Shapiro coined the word "zerotree" to describe a more efficient way to encode the positions of significant wavelet coefficients (coefficients above some threshold). Encoding the values of coefficients isn't enough; you have to know to what part of an image each coefficient corresponds. At high compression rates, recording those positions can eat up a large part of the available "bit budget."

In compression, Shapiro points out, the bits available for encoding an image should be used to encode information one can't predict. "Things that are likely should be cheap."

"A lot of people had tried to predict what is happening at fine coefficients from what is happening at coarse coefficients. I couldn't even get the sign right," he says. "What you can do is predict *lack* of information: if nothing is going on at low frequencies, you can predict that nothing is going on at high frequencies."

His solution was to consider a coefficient at a coarse scale to be a *zerotree root* if it is zero, if all its "descendants" (finer-scale coefficients at the same location) are zero, and if the "root" is not itself the descendant of another zerotree root. ("Zero" here means "insignificant": smaller than some threshold.) Each zerotree root is then encoded with a symbol indicating that its descendants are insignificant. This simplifies encoding the positions of coefficients, since all one has to keep track of are the positions of zerotree roots, the positions of isolated zeros (insignificant coefficients that have at least one "significant descendant"), and the positions of the significant coefficients themselves.

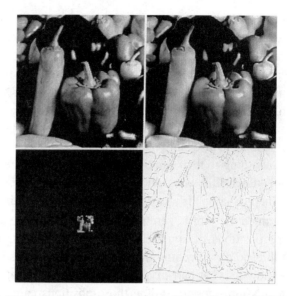

Figure 5. Compact Encoding from Edges, Using Wavelets. After wavelets were used to compute the edges of the image at different resolutions, an algorithm selected the most important ones—in this case, the longest—producing a compression factor of 40. The original image is at top left, the encoded edges at bottom right, and the reconstructed image at top right. (Courtesy of Stéphane Mallat and Sifen Zhong.)

Quantization and the Choice of Wavelet

Another factor influencing the effectiveness of wavelet compression is the choice of quantizer. The most familiar form of quantization is "rounding off" to the nearest whole number. More sophisticated quantizers take into account what is known about human perception, using more bits to encode information that is important for human perception (the edges in pictures, for example), and skimping on information that is less important, just as someone trying to make their house presentable just before a dinner party will pick newspapers and toys up off the floor but will not start organizing bedroom closets.

The wavelet used is also important. In image compression, it is thought that symmetric wavelets make quantization errors less noticeable—that, as Daubechies writes, "it is a property of our visual system that we

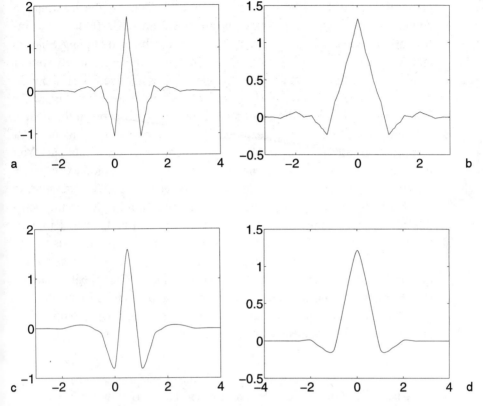

Figure 6. Biorthogonal Wavelets Often Used in Image Compression. (a) wavelet used for decomposition; (b) its scaling function; (c) wavelet used for reconstruction; (d) its scaling function. All were calculated using filters computed by Albert Cohen, Ingrid Daubechies and J. -C. Feauveau. (Courtesy of Stéphane Mallat.)

are more tolerant of symmetric errors than asymmetric ones" [Daubechies, p. 254]. While Daubechies wavelets are asymmetrical, it is possible to create symmetric wavelets with compact support by using two sets of wavelets, one to decompose the signal and another to reconstruct it; such wavelets are called *biorthogonal* (for some examples, see Figure 6).

Using biorthogonal wavelets and a particular form of quantization called vector quantization, Michel Barlaud at the University of Nice-Sophia Antipolis (see [Antonini et al.], [Barlaud et al.]) has achieved compression rates on the order of 50 to 100. That is, the information content of the compressed image is about 1/50th or 1/100th of the information content of the original. Other researchers, including Shapiro and Michael T. Orchard of Princeton University, have achieved similar results. According to Daubechies, image compression factors of about 35 or 40 have been achieved with wavelets, with little loss (see Figure 5). This corresponds to a little less than .25 bits for each pixel in the image.

But compression can't be judged by compression rates alone. "A compression rate in itself doesn't mean much," Barlaud points out. You need to know what image is encoded, the quality and complexity of the original image, as well as the quality of the encoding. Even comparing two methods applied to the same image can be difficult. "Standard methods of objective comparison exist—and they are all bad," says Rioul, at the Ecole Nationale Supérieure des Télécommunications in Paris. How can one objectively compare the quality of a compressed image, when quality is subjective, in the eyes of the beholder? Complexity matters, too. For commercial use, the price of computations has to be affordable.

The final word on the contribution of wavelets to image compression isn't in. In 1992, Daubechies expressed some reservations:

> If you just buy a commercially available image compressor you can get a factor of 10 to 12, so we're doing better than that. However, people in research groups who fine tune the Fourier transform techniques in commercial image compressors claim they can also do something on the order of 35. So it's not really clear that we can beat the existing techniques. I do not think that image compression—for instance, television image compression—is really the place where wavelets will have the greatest impact.

Others are more optimistic. Recently, Kannan Ramchandran and his former student Scott LoPresto at the University of Illinois at Urbana-Champaign and Michael Orchard at Princeton developed what they call the Estimation Quantization (EQ) algorithm, with potential applications in image and video compression. At medium to high compression rates (at or below about 0.25 bits per pixel), "our preliminary results appear to outperform all reported results in the wavelet image coding literature, to the best of our knowledge," they write [LoPresto et al., p. 1].

Like Shapiro's embedded zerotree wavelet coder, the EQ coder takes advantage of the fact that significant wavelet coefficients tend to be clustered together rather than randomly distributed. However, unlike Shapiro's coder, it does not use a zerotree-like structure. In the EQ coder, "wavelet coefficients are modeled as being random-valued with a certain degree of 'uncertainty,'" Ramchandran says.

In a simple code like the Morse code developed for telegraph operators, both the person sending the code and the person receiving it agree to a set of straightforward rules: a single dot means "e", for example, while "dot, space, dot" means "z." In the EQ coder, both encoder and decoder agree ahead of time on the rules of the game, but these rules involve the probability that a given wavelet coefficient will be similar to nearby coefficients, based on empirical observation of "typical" wavelet image data.

The general idea is that since both encoder and decoder agree on the rules, there's no need for the encoder to send information that the decoder can figure out for itself. In transmitting English text, for example, the rules of the game might include the fact that the letter "q" is usually followed by "u." Then only when the encoder runs into something that doesn't obey the predictions (the word "Iraq," perhaps, or "Qatar") does it have to tell the decoder what to put after a "q."

This analogy isn't exact, because the predictions on which the EQ coder is based are not predictions as to the values of wavelet coefficients, but predictions about how uncertain they are. For example, if the value of a coefficient is expected to be a whole number somewhere between -2 and +2, then it is less uncertain than a coefficient that is expected to be a whole number somewhere between -100 and +100. In the first case, the encoder needs to send only enough information to tell which of five possible values is the right one; in the second, it has to send enough information to specify which of 201 possible values is right. In either case, the decoder can interpret that information correctly because it can figure out, from the rules of the game, what the expected spread of values is. (Basically, in a region of high activity (edges, for example), the "uncertainty" attached to a given coefficient is high, while in a region of low activity the uncertainty is low.) The decoder can also deduce how much precision was lost during quantization for each coefficient.

Occasionally this system doesn't work. "The rules of the game are such that the decoder can't reliably predict uncertainties in some places," Ramchandran says. In regions of low activity, the rules will predict a small range for the uncertainty; the result will be that one coefficient after another will be quantized to zero, in a kind of domino effect. In those situations, he says, "the rules of the game call for the decoder to not try to be too smart, and look instead for help from the encoder."

Wavelets are also being applied to compression of three-dimensional data, such as some data from medical imaging. Lars Lippert and Markus Gross at the Federal Institute of Technology in Zurich have devised an algorithm [Lippert et al.] for distributing and rendering huge three-dimensional data sets over the Internet or other networks, where a remote server maintains data sets that are inspected, browsed through, and rendered interactively by a local client. "A very important aspect of our setup is that the client does not need to provide storage for the data," Lippert says. Data are transmitted progressively; if the transmission is interrupted, "the scheme guarantees that the maximum possible energy of the signal is transmitted."

In addition, a number of researchers are exploring the use of wavelets in medical tests, exploiting their ability to concentrate the information of an image in a relatively small number of coefficients. In magnetic resonance imaging, Weaver and Healy [Weaver, Healy] were able to follow the outline of the beating heart by sampling a few wavelet coefficients. Other researchers have used wavelets to compress digital mammograms [Lucier et al.]. But Michael Unser, formerly at the National Institutes of Health and now at the Swiss Federal Institute of Technology, and Akram Aldroubi, at N.I.H., caution that medical use of lossy image compression, which sacrifices some quality in order to get good compression rates, "is still controversial. One of the problems is that the usual, nonmedical image quality criteria are not necessarily very meaningful in the context of diagnostic radiology. The use of these techniques also raises some delicate legal issues" [Unser, Aldroubi, p. 629].

JPEG 2000

One test of wavelet image compression will be whether wavelets are adopted in some form in the "JPEG 2000" standard for image compression. This new standard will replace the current one, adopted in 1990 by the Joint Photographic Experts Group, a joint committee of the International Standards Organization and the International Telecommunications Union–Telecommunications Division. (For a comparison of image compression with JPEG and with an embedded wavelet code, see Figure 7.) The committee is currently gathering candidate algorithms and architectural frameworks to use in composing a single standard, scheduled for adoption in the year 2000.

According to Richard Clark, editor of two JPEG standards and webmaster at www.jpeg.org, the primary goal is good quality at high compression rates (less than 0.25 bits per pixel for highly detailed grey-scale images). But the wish list is long, including progressive transmission; the ability of a decoder to run in real-time through channels with limited bandwidth; "random codestream access and processing," in which the user can access or decompress with less distortion than the rest of the image those regions that particularly interest him; robustness to transmission errors; an open architecture that works well for different types of image and applications; protection of digital images; compatibility with current JPEG standards; and an interface allowing the interchange and the integration of the still-image JPEG 2000 tools and those used by MPEG-4 for moving pictures and audio.

Computational Short-Cuts

Another application of wavelets that may prove very important is the compression of huge matrices (square or rectangular arrays of numbers) in the hopes of solving nonlinear partial differential equations. These techniques were developed by Gregory Beylkin at the University of Colorado at Boulder and Ronald Coifman and Vladimir Rokhlin, both at Yale. The matrix is treated like a picture to be compressed; when it is decomposed into

Figure 7. Comparison of JPEG and an Embedded Wavelet Code. (a) Original Lenna; (b) JPEG compression at .25 bits per pixel; (c) embedded wavelet compression at .25 bits per pixel; (d) original Barbara; (e) JPEG compression at .25 bits per pixel; (f) embedded wavelet compression at .25 bits per pixel. (Courtesy of Stéphane Mallat.)

wavelets with five or six vanishing moments, "every part of the matrix that could be well represented by low-degree polynomials will have very small coefficients—it more or less disappears," Beylkin says. For most matrices with n^2 coefficients, almost any computation (such as a multiplication or an inversion) requires at least n^2 calculations, and often as many as n^3. With wavelets, and for a certain class of matrices, one can get by with n calculations—a very big difference when n is large.

Talking about numbers "more or less" disappearing, or treating very small coefficients as zero, may sound sloppy, but it is "very powerful, very important"—and must be done very carefully, Grossmann says. It works only for a particular large class of matrices: "If you have no *a priori* knowledge about your matrix, if you just blindly use one of those things, you can expect complete catastrophe."

Just how important these techniques will prove to be is still up in the air. In 1992, Daubechies predicted that "five, certainly ten years from now you'll be able to buy software packages that use wavelets for doing big computations, in simulations, in solving partial differential equations." Coifman is optimistic as well. Meyer is more guarded: "I'm not saying that algorithmic compression by wavelets is a dead end, on the contrary I think it's a very important subject. But so far there is very little progress, it's just starting." Of the matrices used in turbulence, he says, only one in 10 belongs to the class for which Beylkin's algorithm works. "In fact Rokhlin has abandoned wavelet techniques in favor of methods adapted to the particular problem; he thinks that every problem requires an ad hoc solution. If he is right, then Ingrid Daubechies is wrong, because there won't be 'prefabricated' software that can be applied to a whole range of problems, the way prefabricated doors or windows are used in housing construction."

But the two approaches—taking full advantage of a particular problem and trying to solve a large class—are not exclusive, Beylkin says. "If the problem is important enough and has many applications, then this point of view is dominant in applications anyway. There is no contradiction with the desire to solve a class of problems (perhaps losing a little bit of speed, say a factor of two to three). Wavelets tend to provide reasonably good solutions to a wide class of problems. Both approaches are needed."

The issue is complicated by the ambiguity of the word *wavelets*. "If it is understood strictly to mean objects generated by quadrature mirror filters (the "classical" wavelets of the Mallat-Meyer multiresolution theory), "then the applicability of wavelets is sharply restricted," Rokhlin says. There do exist classes of problems where wavelets are useful, although not optimal; "there also exist large classes of problems for which wavelets are completely useless, while other structures (localized cosine transforms, for example) work very well."

"On the other hand," he points out, "there exist many 'wavelet-like' bases that are not wavelets in the strict sense, are very similar to classical wavelets, and are in many cases far preferable to the latter. If we agree that these bases are also wavelets, the applicability of wavelets is extended significantly."[6]

In addition, even if it is eventually possible to create easy-to-use software packages incorporating wavelets, wavelets would be only part of the solution. As in the case of image compression, in this setting wavelets can't just be dropped into an existing scheme the way one can open up a computer to stick in some extra memory. Beylkin and his former student James Keiser have devised a spatially adaptive scheme for nonlinear partial differential equations in 1+1 dimensions (time plus one spatial dimension) [Beylkin, Keiser]. "Extending this to higher dimensions requires careful treatment of boundary conditions, and this, in the context of adaptive schemes, is a difficult problem," he said in 1997. "I am quite sure that it will be solved, but, by that time, wavelet ideas will represent only a portion of the total innovation—as is already the case in lower dimensions."

Wavelets and Turbulence

Rokhlin works on turbulent flows in connection with aerodynamics; Marie Farge in Paris, who works in turbulence in connection with weather prediction, remains confident that wavelets will prove to be an effective tool.

---

[6]See for example [Alpert], [Alpert et al.], and [Gortler et al.]. "Wavelet-like bases" does not refer to localized cosine transforms, which are not wavelet-like and work, when they do, for different reasons.

She was working on her doctoral thesis when she heard about wavelets from Alex Grossmann in 1984. "I was very much excited—in turbulence we have needed a tool like wavelets for a long time," she said. (Later she learned that researchers in turbulence in the former Soviet Union, in Perm, had worked independently with similar techniques since 1976.)

> When you look at turbulence in Fourier space, you see cascades of energy, where energy is transferred from one wave number [spatial frequency] to another. But you don't see how those cascades relate to what is happening in physical space; we had no tool that let us see both sides at once, so we could say, *voilà*, this cascade corresponds to this interaction. So when Alex showed me that wavelets were objects that allowed one to unfold the representation both in physical space and in scale, I said to myself, this is it, now we're going to get somewhere. I invited him to speak in a seminar and told everyone in the turbulence community in Paris to come.

> I was shocked by their reaction to his talk. 'Don't waste your time on that,' they told me. 'Nothing proves that it's going to work. Finish your thesis.' Now some of the people who were the most skeptical are completely infatuated, and insist that everyone should use wavelets. It's just as ridiculous. It's a new tool, and one cannot apply it blindly to problems as shapeless as turbulence if it isn't calibrated first on academic signals that we know very well. We have to do a lot of experiments, get a lot of practice, develop methods, develop representations.

Prehistoric Zoology

Farge compares the current state of turbulence research to "pre-scientific zoology." Many observations are needed, she says, to see what structures in turbulence are dynamically important and to try to recreate the theory in terms of their interactions. Possible candidates are ill-defined creatures called "coherent structures" (a tornado, for example, or the vortex that forms when you drain the bath). Farge uses wavelets to isolate them, to try to see how many exist at different scales or whether a single structure exists at a whole range of scales.

Identifying the dynamically important structures would tell researchers "where we should invest lots of calculations and where we can skimp," Farge said. This is important because studying turbulence requires calculations that defy the most powerful computers. The Reynolds number for interactions of the atmosphere—a measure of its turbulence—ranges from 10^{10} to 10^{12}; computer simulations of turbulence based on the fundamental Navier-Stokes equations can now handle Reynolds numbers on the order of 10^2 or 10^3.

So far results have been disappointing, Meyer says. "In turbulence there are phenomena at very different scales, and everyone who studies turbulence thinks that wavelets are the ideal tool for studying the interactions between these scales. It's bizarre, but so far it just doesn't work; no one has a real scientific fact to offer." Certainly wavelets do not provide a straightforward recipe for solving nonlinear equations (such as the Navier-Stokes equation used to describe turbulent flows) in the way that Fourier turned complicated linear equations into cookbook problems.

Beyond Plain English 16

Which Wavelet?

An advantage—and disadvantage—of wavelets is that they are a general tool. Yet there are infinitely many different wavelets and wavelet-Fourier hybrids. So far, there is no consensus as to how hard one should work to choose the best wavelet for a given application, and no firm guidelines about how to make such a choice.
An obvious first decision is the type of representation (continuous or discrete? frame, orthogonal transform, biorthogonal transform?). If you choose an orthogonal transform, you still face an infinite number of choices. How many vanishing moments should your wavelet have? How regular should it be, and how selective in frequency? None of these properties is free. The more regular a wavelet is, or the more vanishing moments it has, the more complex the computations, for example. Some progress has been made on deciding what properties are important for what applications. For example, using wavelets with five or six vanishing moments makes sense for numerical analysis, but not for image encoding. But it is still largely an open question.

See p. 239.

Fourier analysis is not at all suited to nonlinear situations like those found in turbulence, Meyer says. "Wavelets are structurally a little better adapted to them. But is something that is better in principle actually better in practice?" He is disturbed that when wavelets are used in nonlinear problems they

> are used in a neutral way; they are always the same wavelets—they aren't adapted to the problem.... It is in this sense that there is perhaps a doubt about using wavelets to solve nonlinear problems. Scientists always think that there is no miracle solution that works for all problems. What can one hope for from methods that don't take the particular problem into account? At the same time, there are general methods in science which, when one has worked a great deal on particular situations, give some coherence, an overall view. So one can give a different answer depending on one's personality.

Sensual or Austere: Continuous vs. Discrete

Farge uses orthogonal wavelets, or related wavelet packets, for compression, but she uses the continuous wavelet transform for analysis. With a continuous transform she can take a signal that varies with space and see at first glance, on a computer screen or on a printout, what is happening at different scales at any point of the signal. "I would never read the coefficients themselves in an orthogonal basis, they are too hard to read," she says.

"Discrete wavelets are like base 2 or base 10," says Meyer, "while continuous wavelets are like a film of reality, where nothing is lost." Discrete wavelets contain the same information, since one can always reconstruct the signal, but in the interest of efficient algorithms "one loses the intuition, there something is irremediably lost."

Once, he adds, Jean Jacques Rousseau invented a musical notation based on numbers rather than notes on a staff, which would no doubt have been much easier to write than a musical score, only to be told that it would never catch on, that musicians wanted to see the shape and movement of music on the page. This preference isn't limited to musicians. "No one likes successions of points, they are jarring," Meyer says. "Physicists, who have a profound and emotional relationship with reality—much more so than mathematicians, whose relationship is more abstract—can't stand digital representations. They like beautiful images made with continuous transforms, that correspond to something they can grasp, can touch."

Others would argue that mathematicians are every bit as attached to pictures as their cousins. If they prove theorems with formulas, many of them think in geometrical images, which they struggle to reproduce on the two meager dimensions of a blackboard or piece of paper. "All mathematics is sensual," one mathematician said to a student who wanted his mathematics "austere." Still others object that the discrete is not austere. Computers, Morlet jokes, will have taught the young to see our continuous world as a collection of discrete points, like so many pixels.

In addition to aesthetic sensibilities and the ease with which one can read a transform, there's the question of the price one is willing to pay. The point of view of those who work with continuous transforms, like Farge or researchers at Marseille, "comes from their background as experimental physicists," says Meyer. "They have a very, very expensive experiment, and if they spend three weeks, or even three months analyzing it, rather than three days, it doesn't matter. Complex valued wavelets provide an analysis that is just a little bit more precise, a little bit finer, on the condition that they use continuous wavelets, which increases the computing time. In what I do with Coifman at Yale, we have the opposite point of view—concision, the greatest possible economy. These two different directions correspond to different optics."

Beyond Wavelets

One contribution of wavelets, Farge says, is that they have "forced people to think about what the Fourier transform is, forced them to think that when they choose a type of analysis they are in fact mixing the signal and the function used for the analysis. Often when people use the same technique for several scientific generations, they become blind to it. A new tool like wavelets forces us to reconsider the problem from scratch."

Beyond Plain English 17

Different Transforms: A Summary

"A very common pitfall when using any kind of transform," writes Marie Farge, "is to forget the presence of the analyzing function in the transformed field, which may lead to severe misinterpretations, the structure of the analyzing function being interpreted as characteristic of the phenomena under study." A chart summarizes the main characteristics of the different transforms discussed in this book— the Fourier transform, windowed Fourier transform, wavelet transform, Malvar wavelets, wavelet packets, and Matching Pursuit.

See p. 249.

Our choice of a system of representation guides what we see, wrote David Marr: "...any particular representation makes certain information explicit at the expense of information that is pushed into the background and may be quite hard to recover" [Marr, p. 21]. He gave the example of numerical bases. It's immediately obvious whether a number in base 10 can be divided by five, or by 10. Determining whether a number can be divided by seven is harder; if the number is big, it's necessary to make some computations, and computations introduce the possibility of mistakes just as one inevitably makes mistakes—round-off errors if nothing else—when computing the phases of a Fourier transform in order to get time information.

As work with wavelets proceeded, it became clear that if Fourier analysis has limits, wavelet analysis does also. Fourier analysis is suited to very regular periodic signals, while wavelets are suited to highly nonstationary signals with sudden peaks or discontinuities. For quasi-stationary signals, whose behavior is predictable for a certain time, one would want to combine features of the two; if for a long time a signal oscillates steadily, doing nothing surprising or violent, then it's not reasonable to analyze it with little wavelets that only catch a few oscillations at a time. And wavelets are imprecise about frequency in the high frequencies, compared to windowed Fourier analysis. For these reasons, Meyer says, wavelets aren't well adapted to music and speech. Yet standard windowed Fourier analysis is incompatible with orthogonality and, because of its fixed window size, lacks the flexibility of wavelets.

So he and some of his colleagues took a step backwards with a new point of view inspired by wavelets, seeking to create a system of representation that would be orthogonal (and thus suited to fast algorithms) yet still offer the selectivity in frequency of windowed Fourier analysis. They created two.

Wavelet Packets

The first such "hybrid"—*wavelet packets*—was constructed in the summer of 1989 by Meyer and Ronald Coifman working in Gstaad, Switzerland, and Victor Wickerhauser and Stephen Quake, working at Yale [Coifman et al.]. Loosely speaking, a wavelet packet is the product of a wavelet and a wiggle, an oscillating function. The wavelet itself reacts to abrupt changes, while the wiggle inside reacts to regular oscillations. "The idea is to introduce a new freedom," Meyer said. The "window" size, frequency, and position can all be varied independently, as in a musical note, as opposed to wavelets, where all the "high notes" are short (small, high-frequency wavelets) and all the "low notes" are long (big, low-frequency wavelets). Since the choice of wiggles is infinite, "it gives a family that is very rich. It is much more complex, much more subtle, than windowed Fourier," Meyer says. In addition, it's possible to use efficient algorithms.

Farge uses wavelet packets in her work on turbulence, Healy and Weaver in their medical applications. (For a discussion of their use in image compression, see [Wickerhauser 1].) But these functions have in a sense gotten ahead of theory; researchers like Albert Cohen at the University of Paris-Dauphine are working on how to interpret their coefficients, which isn't always straightforward. "We have an enormous mathematical literature to help us interpret wavelet coefficients," says Meyer. "Now that we have understood that the wavelet transform repeats a certain chapter of mathematics, we at least have a science behind us. Wavelet packets are so new that we don't know yet how to interpret the coefficients."

Malvar Wavelets

Coifman and Meyer created their second family of hybrids the following summer during the International Congress of Mathematicians in Kyoto. Since windowed Fourier analysis with a Gaussian envelope cannot be orthogonal, Coifman and Meyer kept the idea of a window filled with trigonometric functions but played with both the shape of the window and the functions that fill it. After several days of discussions, they sat down one afternoon on a bench in the gardens of the imperial palace, while the others at the congress took part in a sightseeing excursion. They ended by constructing a function with a very special shape that begins with an attack, then stretches out in a plateau, and ends with a decrescendo, as shown in Figure 1. This function, which Meyer calls a *Malvar wavelet*, is filled with either sines or cosines but not both (for more details, see [Meyer 3, pp. 75–87]).

Unlike wavelets and wavelet packets, which are essentially artificial, these Malvar wavelets have a physical meaning, Meyer says. "It's really the idea of the emission of a sound for a certain time": a musical note with an attack, a stationary period, and a dying away. As with wavelets, one can vary the size of the window, but this variation is more flexible. Not only can the size of the window be changed independently of the

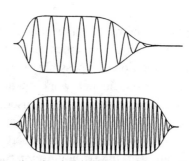

Figure 1. Examples of Malvar Wavelets. A Malvar wavelet consists of an attack, a stationary period, and a dying away. Malvar wavelets make it possible to cut a signal in a nonuniform way: the length of the attack, the length of the stationary period, and the length of the decay can vary independently. Thus they are especially suited to signals like music or speech, where one is primarily interested in the dynamics of the signal as a function of time.

number of oscillations inside, but one can vary independently the length of the attack, the length of the stationary period, and the length of the decay. This possibility of cutting up a signal in a nonuniform way should be useful, Meyer says, for signals like music and speech where one "is interested first of all in the dynamics of the signal as a function of time, and only secondarily in the global frequency content."

Beyond Plain English 18

Wavelets, Music, and Speech

While Malvar wavelets show promise in applications to speech and music, and windowed Fourier analysis is a fundamental tool in speech and acoustics, classical wavelets have not had a strong impact on those fields—partially, says one French researcher, because "the same kind of techniques, without the name 'wavelet,' already long existed in these fields."

See p. 253.

The segmentation takes place automatically, the algorithm choosing the most concise encoding. That this criterion of concision is the right one for encoding speech is "an act of faith, but experimentally it works fairly well," Meyer says. "People who really do phonetics would find this automatic work very crude compared to what they can do with much subtler techniques, but it gives a first classification." Wickerhauser, with Eva Wesfried, Christophe d'Alessandro and Xavier Rodet, has used Malvar wavelets to separate speech into voiced and unvoiced parts. "I think we can get equally good segmentations into smaller speech units like phonemes," he says.

With three independent variables, Malvar wavelets (like wavelet packets) form a highly redundant system that provides an infinite number of orthogonal bases to choose from, with all the associated efficiency. But speed is obtained at the cost of restricting possible positions to a grid. That's the limitation of existing algorithms, Meyer says. "If you want to detect a physical phenomenon, you have to be as continuous as possible, and not privilege points on a grid compared to intermediary points; but if you want effective algorithms you have to be discrete."

Meyer named these functions in honor of Henrique Malvar, an engineer in Brasilia who had earlier constructed a special case of the general construction, without adaptive segmentation. Meyer also calls them "time-frequency" wavelets to distinguish them from the classical "time-scale" wavelets. Others object that Malvar wavelets aren't wavelets at all, but rather "adaptive windowed Fourier analysis."

"Wavelets have gone off in many directions; it becomes a bit of a scholastic question, what you call wavelets," Grossmann says. "Some of the most interesting very recent things would technically not be called wavelets, the scale is introduced in a somewhat different way—but who cares?" But a novice may get lost in this still very confused terminology. Pitfalls abound. A "large-scale" map magnifies little details, showing small towns and dirt roads, but often wavelet researchers mean the opposite: at a large scale they look at long-lived components of a signal, and at a small scale they study small details.

Even using the same word *wavelets* to describe orthogonal wavelets as well as the functions used in the continuous wavelet transform and those used in discrete, nonorthogonal transforms may be misleading. "As far as I can see, these are all called wavelets just because the same people were working on them," says mathematician John Hubbard of Cornell University. "It's rather like biologists who stuck in the phylum of worms everything they couldn't place anywhere else. Orthogonal wavelets are a special case of biorthogonal wavelets, but aside from that, these various functions are just not part of the same genus."

Even the notions of translation and dilation originally thought to be intrinsic to any definition of "wavelet" have been challenged; the *lifting* scheme introduced by Wim Sweldens of AT&T Bell Laboratories and the Catholic University of Louvain in Belgium [Sweldens] produces "second generation" biorthogonal wavelets not formed by translations and dilations. What then is left? Sweldens suggests thinking of wavelets as "building blocks that can quickly decorrelate data"—an umbrella broad enough to cover all possible "wavelets," at the cost of removing all mathematical precision.

Beyond Plain English 19

The Lifting Scheme

Although "wavelet" was always quite an elastic term, encompassing the wavelets of both orthogonal and continuous transforms, such wavelets share at least something in common: they are formed by translations and dilations of a mother wavelet. The lifting scheme produces "second-generation" wavelets that do not rely on translations and dilations. Such wavelets can be used in situations for which traditional wavelets are ill-suited, for example, analyzing data that naturally occurs on the surface of a sphere.

See p. 257.

Best Basis: Choosing the Right Screwdriver

Malvar wavelets and wavelet packets can be used in the framework of a compression algorithm developed by Coifman, Meyer, and Wickerhauser: Best Basis. The idea, due to Coifman, is to represent each signal in the basis that encodes it with optimal concision. Best Basis plays with the two faces of information, its dual nature, seeking the alloy of time and frequency that will express the signal with the fewest possible significant coefficients.

What one wants, says Mallat, is the representation in which "practically all the coefficients are zero, except a few, which are very big. The function, instead of spreading itself out among all the elements, is very well concentrated." When this happens it's said that there is little "entropy," or disorder.

Since all the bases used in Best Basis are orthogonal, they are, on the face of it, equally concise; no redundancy is built into the system of representation, as in a continuous transform. But an orthogonal representation can be redundant for some signals in the sense that one can deduce something about one coefficient from the preceding one. The coefficients are then said to be *correlated*; this correlation comes from the structure of the signal itself.

One should think of this correlation in terms of probabilities. If a signal is thought of as a sequence of symbols or words, then "as the successive symbols are chosen, these choices are, at least from the point of view of the communication system, governed by probabilities; and in fact by probabilities ... which, at any stage of the process, depend upon the preceding choices," explained Weaver [Shannon, Weaver, pp. 10–11]. "Thus, if we are concerned with English speech, and if the last symbol chosen is 'the,' then the probability that the next word be an article, or a verb form other than a verbal, is very small ... After the three words 'in the event' the probability for 'that' as the next word is fairly high, and for 'elephant' as the next word is very low."

Redundancy then can be thought of not as deliberate repetition, like that obtained with a continuous transform, but as "the fraction of the structure of the message which is determined not by the free choice of the sender, but rather by the accepted statistical rules governing the use of the symbols in question," Weaver continued. "... The redundancy of English is just about 50 percent, so that about half of the letters or words we choose in writing or speaking are under our free choice, and about half (although we are not ordinarily aware of it) are really controlled by the statistical structure of the language." This is just about the right amount for crossword puzzles, he added [Shannon, Weaver, p. 14]. If a language "has only 20 percent freedom, then it would be impossible to construct crossword puzzles in such complexity and number as would make the game popular."

Beyond Plain English 20

Best Basis

Choosing a basis in which to decompose a signal means selecting a compromise between time and frequency. This choice can be represented using *Heisenberg boxes* whose height and width represent the interval of time and the range of frequencies. One can think of the elements of an orthogonal basis as Heisenberg boxes paving an idealized time-frequency plane: the boxes covering the plane without overlapping. Given a signal, Best Basis chooses the orthogonal basis in which the signal can be represented with the least possible area, each significant coefficient of the decomposition represented by a Heisenberg box.

See p. 259.

Best Basis seeks to minimize such inherent redundancy by finding the basis that best resembles the signal. The underlying structure of the signal is then already incorporated in the basis, and the signal can be represented succinctly. A curve formed out of two or three sines or cosines can be represented with economy and order by Fourier analysis, for example. But the Fourier transform of a horizontal line with a single discontinuity is a clutter of sines and cosines—verbose and hard to interpret.

It's a matter, Meyer says, of "deciding, before starting to work, what tool is appropriate. To do some delicate electrical job you need a small screwdriver, but for rough carpentry you need a big screwdriver: you choose the tool depending on the job to be done. It's something that didn't exist previously at all, especially as the choice is automatic. That's the algorithm's charm: entropy automatically selects for you the right screwdriver from the toolbox."

Given a signal, Best Basis decides what basis will encode it most efficiently. At one extreme it might send the signal to Fourier analysis (for signals that resemble music, with repeating patterns). At the other extreme it might send it to a wavelet transform (irregular signals, fractals, signals with small but important details). Signals that don't fall clearly into either group are encoded by a hybrid: in one algorithm, Malvar wavelets, in another, wavelet packets. It all goes very fast; the choice of basis "is made based on a limited number of computations at the beginning; this information is kept, since it is used in the final computations. The two take place in parallel," explains Meyer.

Best Basis and Hungarian Dances

In an unusual bit of musical sleuthing, Best Basis has been used to restore a battered recording of Brahms playing his own work, recorded in 1889 on Thomas Edison's original phonograph machine. "Musicians here are quite excited by Brahms' unexpected mode of playing ... this is a real archaeological dig that needs to be carefully unraveled," Coifman says.

The 1889 recording, which includes parts of Brahms' Hungarian Dance Number 1, provided a rare opportunity for musicologists to analyze a major composer's interpretation of his own music—if only enough of the original sound could be salvaged. For a long time this seemed a hopeless endeavor. Recent attempts to convert the original wax cylinder recording to a more modern form failed, and the cylinder, whose fate is now uncertain, was damaged in the process. In 1935, an LP disc had been cut directly from the cylinder; records thought to be direct copies of that recording are owned by the British Library. These were of such poor quality that one musicologist declared that any musical value could be "charitably described as the product of a pathological imagination" [Berger, Nichols, p. 26].

But Jonathan Berger and Charles Nichols of the Yale School of Music report that using wavelet techniques they have succeeded in extracting "enough meaningful musical data for us to challenge this long held view" [Berger, Nichols, p. 28].

The two worked with Coifman on a cassette made from an LP made from a British Library recording—the music so submerged in noise that "most educated listeners fail to recognize the fact that a piano is playing,"

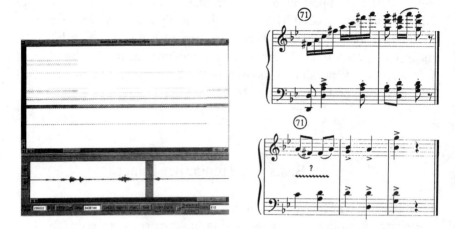

Figure 2. Left, measures 71–72 of the reconstructed and denoised segment of Brahms playing Hungarian Dance Number 1. Right, the score for measures 71–72. The original score at top includes an arpeggio of 16th notes; the score at bottom shows what Brahms actually played, according to the reconstruction. For more information and sound samples, select "research abstracts" on the Yale Music School home page: http://www.music.yale.edu. (Courtesy of Jonathan Berger.)

Berger remarks. Results so far aren't musically pleasing, but by comparing their reconstructed version of the music to the score and to a modern recording, Berger and Nichols arrived at some surprising conclusions.

Brahms took considerable liberties with his own score, doubling the length of eighth notes in some places and shifting the emphasis to the second beat of the measure in others; the performance even lapses into improvisation at several points, as illustrated by Figure 2. The project is continuing. "The result is really a pastiche of many wavelets, many coefficients, many window sizes, and much averaging," said Berger. "Our goal to automate this has quite a ways to go."

The underlying idea was to say that noise can be defined as everything that is not well structured, and that "well structured" means easily expressed, with very few terms, with something like the Best Basis algorithm. So the approach was to use Best Basis to decompose the signal,

Figure 3. Best Basis applied to the problem of deciphering animal language.
(© Hergé/Casterman.)

and remove anything left over. Meyer speculates that this approach might also work with signals such as the speech of dolphins, where researchers don't know what they are looking for. The speech of dolphins must contain a structure in order for dolphins to be able to communicate. Best Basis might help elucidate that hidden structure.

This kind of problem lies outside the traditional field of denoising. "The great things about wavelets is that they let us clearly see that the older theoretical questions are basically settled and they point, for example via wavelet packets, to completely new and stimulating questions," says Donoho. "Now we are trying to use wavelet packets and Coifman-Meyer bases to show statisticians that there is a whole new set of problems to be attacked, where they never looked before."

More recently, Berger (now at Stanford) has been working with Coifman and with Igor Popovic and Maxim Goldberg on a new algorithm that allows for a much more precise positioning of the main threshold. They have also moved beyond Brahms. "Our repertoire of historical recordings that we have worked on has expanded dramatically as we attempt to isolate and organize specific situations into generalizable classes, and to compare our method with other algorithms," Berger says.

Wavelets and Fingerprints

For some purposes, Best Basis is not ideal. Because it "adapts" itself to each signal, it is more complex than nonadaptive algorithms. Compared to one-size-fits-all transforms, it may give superior results when faced with a variety of images, but, for specific applications, a transform custom built for that application will often be best. One such case involves the encoding of the nation's fingerprints.

Since 1924, the FBI has been collecting fingerprint cards, each card containing 14 separate images totaling some 10 megabytes of information. By 1996, the FBI had on file some 200 million fingerprint cards [Brislawn et al.]; each day, from 30,000 to 50,000 new ones arrive. Moving these fingerprints out of filing cabinets into digital files that could be more easily stored, and which would be easily accessible to automated identification systems, was clearly necessary if this bank of information was to be used effectively. Transmitting fingerprints electronically would also be a lot faster than mailing them. But 10 megabytes is a lot of data. With the 9600 baud modems commonly in use in the early 1990s, it would take close to three hours to transmit a single card. (The high-speed modems available in 1997 would bring that down to under an hour, but considering the number of cards to be transmitted, that's still too long.)

Compression was obviously essential in any kind of standardized scheme for fingerprint encoding, and in the early 1990s Tom Hopper of the FBI's Criminal Justice Information Services Division began exploring the possibilities. At that time the Fourier-based JPEG standard was so new that it hadn't even been formally approved by the International Standards Organization. But although it was state of the art, it didn't produce acceptable results: when JPEG is used to compress fingerprints, it destroys fine details and superimposes a blocky pattern on the reconstructed fingerprint.

Coifman and Wickerhauser suggested that wavelets might do better. Indeed, they found they could avoid blockiness using Best Basis and wavelet packets. However, Best Basis turned out to be four or five times slower than two nonadaptive wavelet schemes that produced equally good results: one using vector quantization, developed by Christopher Brislawn and Jonathan Bradley at Los Alamos National Laboratory, and a simpler scheme devised by Hopper, originally dubbed "Plain Vanilla," and intended just as a benchmark by which to measure more complex systems [Hopper, Preston].

The standard adopted by the FBI, based on that Plain Vanilla scheme, took various bits and pieces from existing algorithms and fit them together with a clear focus on a single goal: encoding fingerprints in such a way that the decompressed print is as useful as the original. Far more important than speed, Hopper says, is its ability to reproduce the details of ridge endings and bifurcations—the information that makes it possible to match fingerprints.

Known as WSQ (wavelet scalar quantization), the FBI standard incorporates elements of the Los Alamos algorithm [Bradley et al.], including the way it handles image boundaries; Brislawn and Bradley served as technical advisors on the project. Others in the wavelet field contributed as well. "It was very much a group effort," Hopper says. "I spoke with Brislawn, with Daubechies, with Victor Wickerhauser, with Gil Strang at MIT . . . The people at Aware were helpful also. Everyone was excited to see the FBI embrace state-of-the-art technology."

WSQ, which produces "archival-quality images" at compression ratios of about 15 to 1, uses biorthogonal wavelets devised by Albert Cohen, Daubechies and J.-C. Feauveau (now commonly used in image compression, but at the time very new), and an adaptive form of scalar quantization. Quantization always involves decisions as to what information to keep, and what to throw away. Most image compression algorithms have to be able to handle any kind of image that is thrown at them, which makes quantization something of a balancing act: information important for one type of image may be somewhat less so for another. The FBI has millions of images, all looking roughly alike, so it was possible to zero in on those features that really make one fingerprint different from the next.

The mammoth job of electronic conversion is now complete, Hopper says; by October 1997 some 35 million fingerprint cards existed in digital form, and several states had begun transmitting fingerprints electronically. (For more information on fingerprint compression, also see [Bradley et al.], [Brislawn], [Brislawn et al.], and the web site http://www.c3.lanl.gov/~brislawn.)

The Big Dictionary of Matching Pursuit

Another aspect of Best Basis that can be a blessing or a curse is that it is restricted to orthogonal bases. This makes for fast computations, but limits the choices. Particularly when dealing with highly nonstationary signals—signals that change unpredictably—the libraries of orthogonal bases available to Best Basis may be insufficient. In this case the "best" basis may in fact not be very good.

Mallat and Zhifeng Zhang [Mallat, Zhang] have produced a more flexible system, called *Matching Pursuit*. This algorithm matches each part of the signal with the elementary wavelet that most resembles it, with no requirement of orthogonality. "Instead of trying to globally optimize the match for the entire signal, we're trying to find the right waveform for each feature," Mallat says. "It's as if we're trying to find the best match for each 'word' in the signal, while Best Basis is finding the best match for the whole sentence."

Given a signal, Matching Pursuit searches a dictionary containing elementary waveforms, choosing the one that most closely resembles part of the signal. This "word" is subtracted from the signal, and Matching Pursuit chooses the wave that most closely resembles part of the remaining signal, and so on; an example of a signal decomposed with Matching Pursuit is shown in Figure 4. (To some extent, the dictionaries can produce words on command, modifying waveforms to give a better fit.)

As in the fast wavelet transform, Mallat uses here the idea of iterative subtractions. At each step, the information subtracted from the signal is encoded, and the process is repeated with the remainder. But while the fast wavelet transform treats all signals the same way, according to rigid mathematical rules, the new algorithm creates an individual solution for each signal.

Paradoxically, this redundant dictionary provides concise encodings of highly nonstationary signals. An orthogonal basis is like a dictionary with few words, Mallat says. As long as you restrict yourself to signals to which the basis is well suited, it works very well, but as soon as you try to say something unusual, you are forced to use a great many words,

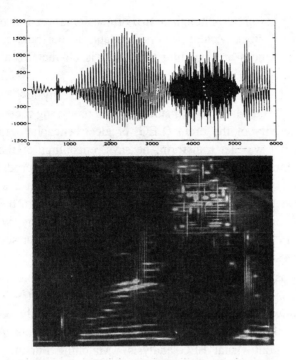

Figure 4. Matching Pursuit, devised for highly nonstationary signals, matches each part of a signal with the elementary wavelet that most resembles it. Above: a speech recording of the word "greasy," sampled at 8 kHz. Below: The time-frequency energy distribution of that speech recording, showing the low-frequency component of the "g" and the quick transition to the "ea." The "ea" has many harmonics, while the "s" has an energy distribution similar to white noise. Most of the signal energy is characterized by a few time-frequency atoms: although the original signal has 3000 samples, it can be reconstructed from only 250 atoms, with a signal to error ratio of 40 db. (Courtesy of Stéphane Mallat.)

to use circumlocutions. With the big dictionary of Matching Pursuit, it's possible to find the right word for each part of a signal.

Matching Pursuit has another advantage. Normally one has to choose between conciseness and translation invariance. But although an encoding by Matching Pursuit is concise, it is translation invariant, since the signal is encoded according to its local characteristics rather than by choosing a starting point arbitrarily and calculating coefficients one after the other. So Matching Pursuit is well suited to pattern recognition. (But not, it appears, to analyzing textures, although a two-dimensional version of Matching Pursuit exists: since Matching Pursuit adapts to the randomness of the texture, it produces an encoding that itself is characterized by random variation.)

Piotr J. Durka of the Laboratory of Medical Physics, Warsaw University, has found Matching Pursuit useful in analyzing recordings of the electrical activity of the brain (EEG, or electroencephalograms). With the orthogonal wavelet transform, he says, encodings of the signal are "neither compact nor elegant. In Matching Pursuit we produce a highly redundant set of functions and then choose from them a small subset that fits best the analyzed signal. In general, choosing a best subset of say n functions for a given signal is a NP-hard problem, which means practically intractable. MP is a sub-optimal algorithm for such a choice, finding a reasonably close solution."

"Funny as it may sound," he adds, "the computer revolution in clinical EEG analysis still consists mainly of the fact that what was previously on paper now is watched at the monitor screen, mainly because most of the new signal processing methods were incompatible with clinical knowledge about visual EEG analysis. Matching Pursuit corresponds directly to the traditional way of analyzing EEG, namely visual analysis. For over 40 years of clinical practice, structures present in EEG were being described in terms like 'cycles per second' or 'time width,' which are directly interpretable in terms of time-frequency parameters. Matching Pursuit is one of the first methods that allow for direct utilization of this huge amount of knowledge collected over years of clinical EEG analysis."

Unfortunately, he says, online clinical applications of Matching Pursuit are still far in the future, because of the massive computations needed. The number of computations depends both on the length of the signal and on the number of waveforms in the dictionary. "The higher resolution we want to obtain, the more waveforms should be included in the dictionary to choose from," Durka says. "For high-resolution analysis of long sequences it can take days even on today's powerful computers."

Matching Pursuit is slower than Best Basis: it takes more time to choose words from a big dictionary than from a small one. While Best Basis encodes an entire signal of n points with $n \log n$ operations, Matching Pursuit takes about that many to find a waveform that matches a single structure of a signal. If one just wants to extract the most important structures (all that is needed for pattern recognition), one can get away with fewer than n^2 computations. The reconstruction, however, is very fast: you just add up the various "words" to reconstruct the original sentence.

Basis Pursuit

More recently, Shaobing Chen and Donoho at Stanford have come up with a kind of Matching Pursuit in reverse, which they call Basis Pursuit. Rather than building up the signal representation waveform by waveform, it begins by decomposing the signal in a basis, and then recursively swaps each waveform of the basis with another waveform chosen from the dictionary, while checking whether this improves the ability of the decomposition to globally approximate the signal with just a few elements of the basis. Each waveform that brings an improvement is kept.

The huge number of calculations this procedure would seem to require is considerably reduced by using the fast linear programming algorithm that the mathematician N. K. Karmarkar had developed to optimize the routing of long distance phone calls for AT&T, saving millions of dollars every year [Karmarkar].

Basis Pursuit addresses a weakness of Matching Pursuit. Because Matching Pursuit is a "greedy" algorithm that chooses each element one after the other, without any overall vision of the ultimate goal, it may come up with a solution that is less than optimal.

Mallat compares a choice of signal decomposition to trying to determine a path to go from one place to another. "If each time you choose the road which seems to go in the best direction you may end up in a dead end or making a big detour. If you plan the whole path in advance by looking at a map, at one point you may choose a road which does not go in the good direction to reach a better road which will lead you quickly where you want to go. This is to say that planning is better, if you have enough time. The same for a global optimization, if you have enough choice."

Like Best Basis, the algorithm of Chen and Donoho makes a global optimization; like Matching Pursuit, it does not restrict itself to orthogonal bases. The result is that it provides a better match than either—at the cost of requiring many more computations.

Future Directions

The quest for new ways to encode information is far from over. New languages with extremely varied possibilities have been invented; now one must learn to exploit these riches. For a given task—video compression, transmitting speech, medical diagnosis, solving equations, to mention only a few—should one use Fourier analysis, wavelets or a hybrid?

If a wavelet transform, should it be orthogonal, biorthogonal, continuous, or a frame? With or without many vanishing moments? Is regularity important? What criteria should one follow? Concision is one, but there are others: speed of computations, ease of interpreting coefficients, quality of the results. Researchers have only begun to answer these questions; a solution that seems optimal today may be outdated tomorrow. And some tasks may require still other forms of representation.

"When you speak, you have a huge dictionary of words and you pick the right words so that you can express yourself in a few words. If your dictionary is too small, you'll need a lot of words to express one idea," Mallat says. Yet if it is too big, you may be paralyzed by an embarrassment of riches, unable to choose a word from a long list of synonyms.

In the ideal world, one would have all the words one needs, without excess. Mallat describes an experiment in which a cat was raised from birth in an environment with only horizontal bars. "After a year, if you show the cat vertical bars, he doesn't see them. Since vertical bars don't exist in his environment, he has no need to see them, but he has an excellent ability to distinguish horizontal bars.

"I think the challenge we are facing right now is, when you have a problem, how are you going to learn the right representation. The Fourier transform is a tool, and the wavelet transform is another tool, but very often when you have complex signals like speech, you want some kind of hybrid scheme. How can we mathematically formalize this problem? How can we find the right representation and compute it fast?"

In some cases the task goes beyond mathematics; the ultimate judge of effective compression of a picture or of speech is the human eye or ear, and developing the right mathematical representation is often intimately linked to human perception. Information is not all equal. Even a young child can draw a recognizable outline of a cat, while the very notion of a drawing without edges is perplexing, like the Cheshire cat in *Alice's Adventures in Wonderland*, who vanishes leaving only his smile. Other differences are less understood. People have no trouble telling textures apart, for example, while "after 20 years of research on texture, we still don't really know what a texture is mathematically," Mallat says.

Wavelets may help with these questions, especially since wavelet-like techniques are used in human vision and hearing, but any illusions researchers may have had that wavelets would solve all problems unsuited to Fourier analysis have long since vanished. As Meyer wrote [Meyer 3, p. 8], "while a single algorithm (Fourier analysis) is appropriate for all stationary signals, the transient signals are so rich and complex that a single analysis method ... cannot serve them all."

It may not be the least of the contributions made by wavelets that they have inspired both a closer and a broader look at mathematical languages for expressing information: a more judicious look at Fourier analysis, which was often used reflexively ("the first thing any engineer does when he gets hold of a function is to Fourier transform it—it's an automatic reaction," one mathematician said) and a more free-ranging look at what else might be possible. "When you have only one way to express yourself, you have limits that you don't appreciate," Donoho says. "When you find a new way, it teaches you that there could be a third or a fourth way. It opens up your eyes to a much broader universe."

PART II
Beyond Plain English

Apologia

The material that follows—in *Beyond Plain English* and in the Appendix—spans an unusual range of levels of mathematical sophistication, with some material properly belonging in a high-school curriculum and other (I am told) more naturally belonging to a second-year graduate course. It is perhaps inevitable that mathematically sophisticated readers will feel at times that they are slumming, or at least that they are in a very strange neighborhood. If you are among them I ask your indulgence. It would be rude and not quite true to say that this book wasn't written for you, but my primary motivation was to write it for people who don't know much mathematics. It has been suggested that it would be more seemly to exclude the elementary material, which everyone over age 15 knows anyway. I haven't done so because it's all material that I either never knew or had forgotten, as I have forgotten how to read the *Aeneid* in Latin. I believe I am not alone, and I don't think it realistic or kind to ask the nonmathematical reader to hunt up an old high-school text in order to read this book. One might as well say that he or she has no right to be interested in mathematics.

Some of the elementary material (a review of trigonometry, for example) is self-contained; if it offends, just avert your eyes. The pace of some more advanced articles may pose more of a problem, but I couldn't have written them any other way, and not just because I want them to be accessible to readers like myself. Writers who try to communicate to nontechnical readers a technical subject in which they themselves are not expert are sometimes tempted to parrot the experts without understanding them. This has disastrous effects. The lay reader sometimes has the illusion that he understands something, but on reflection he realizes that his ideas are vague and amorphous, for which he often blames himself. Soon he abandons any attempt to understand subjects that he judges too difficult.

The experts aren't happy either. Inevitably the writer has to string together their words using his own words; he also has to make the subject dramatic; no one wants to read a pastiche of paragraphs copied from professional journals. When the writer doesn't understand what he wants to say, every word is an invitation to disaster. Seeing their words taken out of context—or words spoken in an excess of enthusiasm taken seriously— the experts cringe at the thought of their colleagues' ridicule and swear that never again will they speak to a journalist.

I like to understand what I write, or at least to think that I understand it. (I cheat sometimes, but not often.) This requires that I go to some trouble to make things easy enough so that I can understand them. Sometimes, on rereading something I wrote weeks before, I discovered that I no longer understood it and had to rewrite it so that I could understand it again. As part of this approach, I have in some articles practiced multiresolutional writing, discussing first in general terms what a formula says, then more specifically, and finally writing the formula itself. Some readers may find this irritating. In reading mathematics there are two extremes. The approach that comes naturally to me is to read the text and skip the formulas, while my husband prefers to read the formulas, looking at the text only as a last resort. If you find the pace of discussion painfully slow at times, I suggest you try his method.

Writing on an elementary level doesn't excuse errors. A mistake in an elementary book is more harmful, I think, than in an advanced book whose readers know enough not to be misled. I will not be so rash as to claim I have made no mistakes, but I have tried to communicate without "straying from the truth." To again borrow an image from wavelet theory, I see this book as giving a low-resolution picture of Fourier analysis and wavelets, but not a distorted one.

For those readers who want a higher-resolution picture, I have included a list of more advanced wavelet books, as well as detailed references. I once read of a survey designed to determine the honesty or dishonesty of scientists; among the damning questions asked was, "Have you ever listed a reference that you did not yourself read, or do you know anyone who ever did so?" Let me plead guilty right away. It should be patently obvious that I have not read, for example, Haar's "Zur Theorie der Orthogonalen Funktionensysteme" or even Strömberg's "A modified Franklin system and higher-order spline systems on R^n as unconditional bases for Hardy spaces." (In at least one case, though, I have read parts of an article cited by some wavelet researchers who never read it, so I am in good company.) If you are a "nonmathematical" reader, I hope you will not follow my lead and skip the formulas. The excuse for skipping formulas is the conviction that they are going to contain some unknown symbol or unfamiliar notation, so that trying to understand them is as pointless as undertaking a 500-piece puzzle knowing in advance that several pieces are missing. I have done my best to avoid this problem. Of course, the articles are of different degrees of difficulty; if you are not at ease with trigonometry, for example, following some of the steps in *Multiresolution* will be difficult. But a reasonable effort to read the major formulas—to see why they correspond to the text—should, I hope, make it possible to follow the gist of the discussion. The proofs given in the Appendix were also written with you in mind, although they necessarily contain steps you will have to take on faith.

You may ask, why bother? Reading this book is not a shortcut to a proper mathematical education; you won't be able to pass an exam at the end or use Fourier analysis or the wavelet transform in any practical way. For the reader who intends to pursue more mathematical studies, it may, I hope, provide some motivation and give an idea of where one is trying to go, emphasizing ideas that may sometimes get lost in the thicket of details and techniques. For the reader who does not plan to study mathematics further, I hope it will give both some insight into what mathematicians do, and some appreciation for the power and beauty of techniques that have had such an impact on our daily lives and on the way we think about the world.

The Fourier Transform

The Fourier transform is the mathematical procedure that breaks up a function into the frequencies that compose it, as a prism breaks up light into colors. It transforms a function f that depends on time into a new function, $\hat{f}$, or "f hat," which depends on frequency. This new function is called the *Fourier transform* of the original function (or, when the original function is periodic, its *Fourier series*). For functions or signals that vary with time—music, for example, or the fluctuations of the stock market—frequency is most often measured in hertz, or cycles per second, illustrated in Figure 1.

Functions can also vary with space. The Fourier transform of a fingerprint might have important values near the "frequency" of 15 ridges per centimeter. (Strictly speaking, frequency is the inverse of time, so for a function that depends on space one often says *wave number*, the inverse of space.)

A function and its Fourier transform are two faces of the same information. The function displays the time information and hides the information about frequencies. The function corresponding to a musical recording shows how the air pressure (produced by sound waves) changes with time, but it doesn't tells us what frequencies—what notes—make up the music. The Fourier transform displays information about frequencies and hides the time information: the Fourier transform of music tells what notes are played, but it is extremely difficult to figure out when they are played.

Nevertheless, the function and its transform both contain all the information of the signal. We can compute the transform from the function and then retrace our steps, reconstructing the function from the transform. (This process is known as *inverting* the transform.)

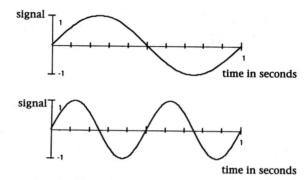

Figure 1. The frequency of $\sin 2\pi x$ is 1 cycle per second (1 hertz). The frequency of $\sin 2\pi 2x$ is 2 cycles per second (2 hertz).

Fourier Series

The Fourier series of a periodic function f of period 1 is written:

$$f(t) = \frac{1}{2}a_0 + (a_1 \cos 2\pi\, t + b_1 \sin 2\pi\, t) + (a_2 \cos 2\pi\, 2t + b_2 \sin 2\pi\, 2t) + \cdots .$$

$$(1.1)$$

The Fourier coefficients $a_1, a_2, a_3, \ldots$ tell how much the function "contains" of the functions $\cos 2\pi t$, $\cos 2\pi 2t$, $\cos 2\pi 3t, \ldots$ (i.e., cosines of frequencies 1 hertz, 2 hertz, 3 hertz...); the coefficients $b_1, b_2, b_3, \ldots$ tell how much it "contains" of the functions $\sin 2\pi t$, $\sin 2\pi 2t$, $\sin 2\pi 3t \ldots$ (i.e., sines of frequencies 1 hertz, 2 hertz, 3 hertz...). A Fourier series concerns only those sines and cosines that are integer multiples of the base frequency.

Equation (1.1) is more commonly written:

$$f(t) = \frac{1}{2}a_0 + \sum_{k=1}^{\infty}(a_k \cos 2\pi kt + b_k \sin 2\pi kt), \qquad (1.2)$$

where k represents frequency, and the "sum" symbol

$$\sum_{k=1}^{\infty}$$

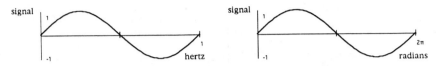

Figure 2. The "2π's" in Fourier analysis make the equations unwieldy but are unfortunately inevitable. We can speak of "sin $2\pi x$ periodic of period 1" or of "sin x periodic of period 2π." The graphs are the same; only the labels vary.

means that one adds together the terms $(a_k \cos 2\pi kt + b_k \sin 2\pi kt)$ for all integer values of k from 1 to infinity. (Usually frequency is represented by the Greek letter corresponding to the variable of the original function: τ, or "tau," the Greek t, for the Fourier transform of a signal that depends on time, and ξ, or "xi," the Greek x, for the Fourier transform of a signal that depends on space. But according to mathematical convention, the letters τ and ξ represent continuous variables, so for Fourier series mathematicians prefer to represent frequency with k, which by convention represents variables that are integers.)

We obtain the coefficients of a Fourier series for a function $f(t)$, periodic of period 1, with the equations:

$$a_k = 2 \int_0^1 f(t) \cos 2\pi kt\, dt \qquad and \qquad b_k = 2 \int_0^1 f(t) \sin 2\pi kt\, dt.$$
(1.3)

(We integrate the product of the function f and the sine or cosine of the frequency k, measuring the area delimited by the new function; the result is multiplied by 2. See *Computing Fourier Coefficients with Integrals*, p. 137, for some examples, and the Appendix, p. 271, for a general discussion of integrals. The symbol $\int_0^1$ means to integrate between 0 and 1; the "dt" at the end says to integrate with respect to t.)

When we write the Fourier series this way, we are measuring the frequency in hertz, i.e., turns per second. As shown in Figure 2, we could also measure the frequency in radians per second. In that case, we would integrate between 0 and 2π, and square roots of 2π would show up in the formulas for the coefficients.

We can reconstruct a function from its Fourier coefficients using equation (1.1): we multiply each sine or cosine by its coefficient to make it

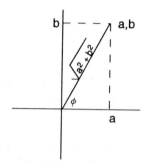

Figure 3. If we represent a function's Fourier coefficients at a particular frequency as the point (a, b) in the plane, the phase for that frequency is the angle ϕ formed by the horizontal axis and the line from the origin to that point, and the amplitude is the length of that line, $\sqrt{a^2 + b^2}$.

taller or shorter, and we add, point by point, the resulting functions; the first term is divided by 2.

Amplitude and Phases

In a Fourier transform, the information on time (or on space) is hidden in the *phases*: the displacement of the sines and cosines for each frequency, affecting how they add or subtract.

If we represent a function's Fourier coefficients at frequency k as the point (a_k, b_k) in the plane, as shown in Figure 3, the phase is the angle ϕ ("phi") formed by the horizontal axis and the line going from the origin to that point. The length of that line, $\sqrt{a_k^2 + b_k^2}$, is the function's *amplitude* at frequency k: the contribution of frequency k to the signal.

The phase ϕ is then related to the coefficients a and b by the following equations:

$$\cos \phi = \frac{a}{\sqrt{a^2 + b^2}}; \qquad \sin \phi = \frac{b}{\sqrt{a^2 + b^2}}.$$

The kth term of the Fourier series is $a_k \cos kx + b_k \sin kx$. (We use here a function periodic of period 2π so that we can talk about sines and cosines of kx rather than $2\pi kx$; ϕ is then measured in radians.) We can write this another way. Multiplying and dividing the kth term of the Fourier series by $\sqrt{a_k^2 + b_k^2}$ gives

$$\sqrt{a_k^2 + b_k^2} \left(\frac{a_k}{\sqrt{a_k^2 + b_k^2}} \cos kx + \frac{b_k}{\sqrt{a_k^2 + b_k^2}} \sin kx \right)$$

$$= \sqrt{a_k^2 + b_k^2} \left(\cos \phi_k \cos kx + \sin \phi_k \sin kx \right)$$

$$= \sqrt{a_k^2 + b_k^2} \left(\cos (kx - \phi_k) \right).$$

(The last step uses formula (B.1) in the review of trigonometry in the Appendix, p. 270.) We see that the kth term of the series can be thought of as a single cosine shifted by the phase angle ϕ_k.

The equation for the phase looks straightforward, but it is virtually impossible to compute phases with enough precision so as to extract time information. The frequency (pitch) of the note A given by a tuning fork, for example, is 440 hertz: 440 cycles a second. If you wanted to determine, from the Fourier transform of the recording of a symphony, whether this A is played 20 minutes into the symphony, you would have to calculate its phase with more precision than 1 over $(20 \times 60 \times 440)$, that is, 1 part in 528,000. (The phase remains unchanged throughout the music, but any round-off error you make by calculating it with less precision gets compounded cycle after cycle.) In order to know where you are in the cycle, you would have to multiply that number by about 5, giving 1 part in 2,640,000. Knowing the phase of just this one frequency is equivalent to measuring a kilometer to within less than half a millimeter.

Another way to think of it is to imagine that the symphony is in A major and therefore its Fourier transform has a big coefficient for that frequency; 20 minutes into the symphony there is a measure of silence. During the measure of silence the A is negated by all the other frequencies. If the phase of that A were changed by just 180 degrees—half a cycle, of which there are 440 per second—then instead of the note A being represented by $\sqrt{a_k^2 + b_k^2} \left(\cos (kx - \phi_k) \right)$, it would be represented by $\sqrt{a_k^2 + b_k^2} \left(- \cos (kx - \phi_k) \right)$ changing the sign of the cosine. Instead of hearing a measure of silence—the frequency A negating the other frequencies—we would hear a very loud A: the "true" A of the symphony plus the sum of all the other notes (a "minus A," which we would hear as the note A).

Fourier Transforms

The only frequencies that contribute to the Fourier series of a periodic function are the integer multiples of the function's base frequency (the base frequency being the inverse of the period). If a function is not periodic but decreases sufficiently fast at infinity so that the area under its graph is finite, it is still possible to describe it as a superposition of sines and cosines—to analyze it in terms of its frequencies. But now we must compute coefficients for all possible frequencies, to compute its Fourier *transform*. The equations are:

$$a(\tau) = \int_{-\infty}^{\infty} f(t) \cos 2\pi\tau t \, dt \quad and \quad b(\tau) = \int_{-\infty}^{\infty} f(t) \sin 2\pi\tau t \, dt$$

$$(1.4)$$

or, by changing the names of the variables:

$$a(\xi) = \int_{-\infty}^{\infty} f(x) \cos 2\pi\xi x \, dx \quad and \quad b(\xi) = \int_{-\infty}^{\infty} f(x) \sin 2\pi\xi x \, dx.$$

(In equation (1.3), we were concerned only with the interval from 0 to 1, since our function was periodic of period 1. Since we are now dealing with a nonperiodic function, we must consider the interval between $-\infty$ and ∞.)

Complex Numbers

Appearances notwithstanding, mathematicians aren't particularly masochistic; if they like to work with complex numbers, it's because it's often easier. In Fourier analysis, complex numbers make it possible to have a single coefficient for each frequency: we no longer need sines and cosines.

A complex number $z = x + iy$ (where i is the imaginary number $\sqrt{-1}$) can be represented by the point (x, y) in the plane. To write a Fourier series or transform using complex numbers, we use the following formula (first published by Euler in 1748), which some mathematicians consider one of the most remarkable and mysterious in all of mathematics, linking two branches of mathematics—trigonometry and the computation of compound interest—that existed entirely independently of each other for more than two thousand years:

$$e^{i\theta} = \cos\theta + i\sin\theta. \tag{1.5}$$

In other words, when we think of θ as the angle theta, the number $e^{i\theta}$ turns eternally around a circle of radius 1. (If this isn't clear, see the brief review of some trigonometric definitions in the Appendix, p. 267.)

Like π, e is irrational, in fact transcendental; like π, it tends to appear in the most unexpected places, like equation (1.5) or in the computation of interest. (If you put a dollar in the bank at 100 percent interest per year, compounded not every day or even every second, but continuously, at the end of a year it would be worth $\$e \approx \$2.718\,281\,853\,9\ldots$.)

In Fourier analysis, we use this mysterious relationship simply as a convenient notation. Equation (1.2) for the Fourier series of a periodic function then is written:

$$f(t) = \sum_{k=-\infty}^{\infty} c_k e^{-2\pi i k t},$$

while the two equations (1.3) for the coefficients of a Fourier series become:

$$c_k = \int_0^1 f(t)\, e^{2\pi i k t}\, dt.$$

(In this notation, the various 2's and 1/2's in equations (1.1-1.3) conveniently disappear.) The term $e^{2\pi i k t}$ is called the *exponential* (more precisely, the *complex exponential*) of frequency k.

The formulas for the Fourier transform of a function that decreases at infinity, and for the reconstruction of the function from the transform, are:

$$\hat{f}(\xi) = \int_{-\infty}^{\infty} f(x)e^{2\pi i \xi x} dx \quad and \quad f(x) = \int_{-\infty}^{\infty} \hat{f}(\xi)e^{-2\pi i \xi x} d\xi.$$

(We are speaking this time of a function that depends on space, so we use ξ rather than τ, and x rather than t.) Other versions of these formulas exist, depending on where one puts the 2π; they are listed in the Appendix, p. 275.

Talking About Functions: f or $f(x)$?

Mathematicians like to differentiate between a function f and its value $f(x)$ at a particular point x. They think of a function f in the abstract; it doesn't matter whether the variable is time, space, temperature, money or whatever, much less what the precise units are. However, it's impossible to define a function without some kind of "place holder" that tells you what it does to its variable. So to define a function f we say, for example, $f(x) = x^2$. This probably contributes to the difficulty many students have with this terminology. Some find it virtually impossible to think of a function f in the abstract; when a professor says f they think $f(x)$.

This is perhaps an example in mathematics where *ontogeny recapitulates phylogeny*. In the nineteenth century, mathematicians always spoke of functions $f(x)$; they only began writing f when they learned to think of functions as points, or vectors, in an infinite-dimensional space (see *Orthogonality and Scalar Products*, p. 153). Yet other concepts that historically gave mathematicians a lot of trouble—complex numbers, for example—are accepted without question by students.

Whatever the cause, this confusion of terminology creates problems. We can correctly say that the Fourier transform of f is $\hat{f}$. But we cannot say that the Fourier transform of $f(x)$ is $\hat{f}(x)$. If f varies with space, then $\hat{f}$ varies with spatial frequency: the Fourier transform of $f(x)$ is $\hat{f}(\xi)$. (We can say that the Fourier transform of $f(x)$ is $\widehat{f(x)}$, but life is complicated enough—and many mathematicians' handwriting bad enough—without having to worry about the width of hats.)

The Convergence of Fourier Series and the Stability of the Solar System

As Fourier showed, virtually any periodic function can be represented as the sum of an infinite number of sines and cosines, each assigned a coefficient. The information that distinguishes one Fourier series from another is contained in the coefficients. For example, the discontinuous square-wave function, periodic of period 2π, can be written as the Fourier series

$$f(x) = \sin x + \frac{1}{3}\sin 3x + \frac{1}{5}\sin 5x + \cdots . \qquad (2.1)$$

For this function, the coefficients are $1, 0, \frac{1}{3}, 0, \frac{1}{5}, 0, \ldots$. (More precisely, these are the coefficients of the sines; the coefficients of the cosines are zero.) Figure 1 shows the square-wave function and the sum of the first seven terms of its series.

To better see how the series *converges* towards the function, Figure 2 shows the sums of the first 1, 2, 3, and 4 terms of the series.

A Kind of Convergence

We had to say that *virtually* any periodic function can be represented as a Fourier series because in 1872, some exceptions were found: continuous functions whose Fourier series diverges from the function at certain points. (Some large-minded mathematicians consider these series convergent, but even these mathematicians boggle at a function discovered by Andrei Kolmogorov in 1923, whose Fourier series diverges everywhere.)

125

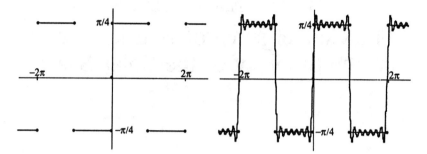

Figure 1. Even a discontinuous function like the square-wave function can be written as a superposition of sines and cosines. Left: the function. Right: the sum of the first 7 terms of its Fourier series. (For this function, the coefficients of the cosines are all zero.)

Actually, even when a Fourier series does converge towards its function, convergence doesn't always work as neatly as one might hope, and the man on the street might find something a bit fishy about the mathematical definition of convergence. A function like the square-wave function, which is *piecewise continuous* (continuous in stretches, between the discontinuities) and *piecewise differentiable* (the parts between the discontinuities are very smooth) has, as Gustave Lejeune Dirichlet proved, a *pointwise convergent* Fourier series.

Within the continuous stretches, excluding the neighborhood at the brink of the discontinuities, the series converges the way one would expect: as one adds more and more terms to the series, the series gets closer and closer to the function, converging to it *uniformly*. Each term added to the series brings it a little closer to the function, for every point in the interval. At the discontinuities this is obviously impossible; in Figure 1, the value of the series at 2π cannot simultaneously converge both to $\frac{\pi}{4}$ and $-\frac{\pi}{4}$. Instead it converges to the average of the two.

In the process, odd things happen, and calling the series convergent requires a bit of mathematical legerdemain. If you look at Figure 2, you will see that the first term of the Fourier series of the square-wave function overshoots the function. As one adds terms, the part that overshoots the function moves closer and closer to the discontinuities, but it doesn't disappear. In fact, the overshoot never disappears and never gets substan-

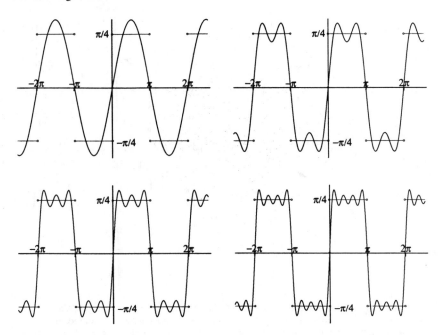

Figure 2. The sum of the first 1, 2, 3, and 4 terms of the Fourier series of the square-wave function. As more terms are added, the partial sum of the series better approximates the function.

tially shorter (approximately one tenth the jump of the function). This *Gibbs' phenomenon* limited the accuracy of mechanical tide predictors built in the 19th century, and it remains a problem today, in medical imaging, for example. But the series still contains all the information of the function, and it is said to be pointwise convergent: at every point the series eventually converges to the function, *just not all the points at the same time.*

One can get rid of the Gibbs' effect by summing the series in a different way. The overshoot is caused by the sharp cutoff of Fourier coefficients. At some point short of infinity one decides that enough is enough and abruptly stops adding terms to the series. The problem disappears if one tapers off the summing of the series, gradually diminishing the proportion of each term that is added. One way to avoid this, described in [Hubbard, McDill] and in the Appendix, p. 291, also provides a much

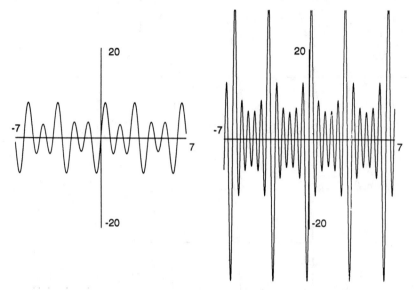

Figure 3. Not all series of sines and cosines represent functions; some, like the series $\sin x + 3\sin 3x + 5\sin 5x + \cdots$, diverge as more terms are added to the series. Left: the sum of the first 3 terms of this series. Right: the sum of the first 6 terms.

easier proof of the convergence of Fourier series than Dirichlet's. But tapering off the summing spoils the efficiency of the algorithm. In addition, with standard Fourier analysis, if you don't like the result, you can always add a few more terms; with tapering, you have to start summing all over again.

Divergence

While virtually every periodic function has a Fourier series, not every series of sines and cosines represents a function. If the coefficients do not become small sufficiently rapidly, the series *diverges* and does not represent a function, like the series

$$\sin x + 3\sin 3x + 5\sin 5x + \cdots , \qquad (2.2)$$

shown in Figure 3.

(In the 1950s, Laurent Schwartz and Israël Gelfand discovered a new approach—*distributions*—that made it possible to give a meaning to some divergent Fourier series, the series (2.2) for example. These methods transformed the theory of linear differential equations.)

It would be difficult to exaggerate the importance of the convergence of Fourier series, for both pure and applied mathematics. It is this problem that foiled attempts by 19th-century mathematicians to prove the stability of the solar system. If one is searching for a periodic or quasi-periodic solution to a problem—for example, whether in a hundred thousand years the planets of the solar system will follow essentially the same orbits as they do today—it is natural to try to represent the solution as a Fourier series. In 1878, Karl Weierstrass wrote the mathematician Sophie Kovalevskaya that he had succeeded in writing Fourier series whose coefficients were determined by the equations of motions of Newton [Weierstrass 1, p. 31], but that he couldn't show that these series converged.

Three years later, still working on the problem, he wrote her [Weierstrass 1, p. 33] that he was more and more convinced that "completely different approaches" would be needed if the problem was ever to be solved: "I catch glimmers of these new paths before me, but always enveloped in mist. If I had someone here, with whom I could talk with every day about everything I am trying, maybe a lot would be clearer to me."

Could Saturn Escape from the Solar System?

If we consider the ultimate destiny of the solar system, the most troublesome case is the couple Jupiter-Saturn, whose years form the ratio 2/5. (In the time it takes Saturn to turn twice around the sun, Jupiter turns five times.) Periodically the two planets find themselves in the same position relative to each other, and one would expect the perturbations due to their mutual attraction to amplify the modifications of their orbits, like the resonance observed when pushing a child on a swing. Even more so: unlike the swing, the movements of the planets are not restrained by friction.

Because of this rational ratio, the physicist Jean-Baptiste Biot predicted that a minute perturbation of the orbit of Saturn or of Jupiter

Figure 4. The great 19th-century mathematician Karl Weierstrass tried to use Fourier analysis to prove that the solar system could remain stable forever. He succeeded in writing Fourier series describing stable orbits, but he was never able to prove the convergence of the series, which contained an infinite number of coefficients with very small divisors. The first proof that a series with small divisors converged was made only in 1942. (United Features Syndicate. © 1992. Reprinted by Permission.)

would throw Saturn out of the solar system. Exasperated, Weierstrass remarked [Weierstrass 2] that Jupiter could escape just as easily, which would "indeed simplify the work of astronomers considerably, since this is precisely the planet causing the largest perturbation." He objected that the stability of orbits could not depend on the rationality or irrationality of the ratio of their orbits. How could one measure orbits with enough precision so that this distinction made sense? Yet these rational ratios are essential in his mathematical description of the problem, and they had to be taken into account.

The problem of the resonance of the two planets is translated mathematically by the notorious problem of small divisors. If two planets whose movements are independent revolve around a sun, then stable orbits forming the ratio 2/5 are perfectly compatible with Newton's equations. But when one takes into account the gravitational force between the two, then the Fourier series describing the orbits has an infinite number of coefficients that contain very small divisors, putting the convergence of the series in doubt. (A coefficient with a small denominator may be very large.)

Weierstrass never managed to prove the convergence of his series. The first proof that a series with small divisors converged (a tour de force by Carl L. Siegel) was made only in 1942 [Siegel].

An Imaginary Bank Account

To understand the relationship between rational numbers, small divisors and convergence, consider the following question. If f is a periodic function, with an average value of 0 over its period, does there exist a function g with the same period, satisfying

$$g(t) - g(t - p) = f(t)?$$
(2.3)

This problem is a linearization of the problem of the orbits of Jupiter and Saturn. Since it is always easier to think about numbers in terms of money, let's consider an imaginary bank account. Our function g represents the amount of money at time t in a bank account in which we make deposits or from which we make withdrawals in a continuous fashion. The function $g(t - p)$ represents the value of this account 24 hours earlier (p, when expressed in years, is approximately 1/365.) The function f, which is given, measures at each instant how much money the account gained or lost during the preceding 24 hours.

Since f is periodic, let's say with a period of one year, the pattern of our withdrawals and deposits repeats every year; if we withdraw a thousand dollars during one 24-hour period, then we will withdraw a thousand dollars in the corresponding period one astronomical year later, or a hundred astronomical years later.

If f is periodic with a period of one year, is it possible to have a bank account whose value $g(t)$ also repeats with a period of one year? Such an account might vary from one moment or day to the next, but its entire history would repeat itself every year, forever. Like Weierstrass, who wanted to know whether the solar system will be stable for all eternity, we want to know whether our bank account is stable forever. Can we be sure never to go bankrupt (fall into the sun) and never become billionaires (escape the solar system)?

An infinite number of bank accounts g exist (none of which is necessarily periodic—that's what we have to determine). We haven't said anything about initial conditions. If we want to know how much money we have in our account at every instant, we need to know how much is in the account at every instant during the first 24 hours; then we can use equation (2.3) to determine g. So we are free to give ourselves as much money as we want during the first day, or rather, we are obliged to consider the infinite family of possibilities. Our question then becomes: out of all the possible bank accounts g, are there any that repeat with a period of one year? Do periodic solutions exist?

Our approach will be the same as that used by Weierstrass and his contemporaries to study the orbits of planets. Can we write g as a Fourier series? Does it converge? We will first find the Fourier coefficients for g. We will then see that the rationality or irrationality of numbers has a decisive influence on whether the series converge—whether they represent an actual solution to our problem.

The first step is surprisingly simple. We rewrite f and g in terms of Fourier series, using complex exponentials:

$$f(t) = \sum_{n=-\infty}^{\infty} a_n e^{2\pi i n t} \quad , \quad g(t) = \sum_{n=-\infty}^{\infty} b_n e^{2\pi i n t}$$

where the a_n (the Fourier coefficients of f) are known and the b_n (the Fourier coefficients of g) are unknown. Equation (2.3) then becomes

$$\sum_{n=-\infty}^{\infty} b_n e^{2\pi i n t} - \sum_{n=-\infty}^{\infty} b_n e^{2\pi i n(t-p)} = \sum_{n=-\infty}^{\infty} a_n e^{2\pi i n t},$$

or:

$$\sum_{n=-\infty}^{\infty} b_n (1 - e^{-2\pi i n p}) e^{2\pi i n t} = \sum_{n=-\infty}^{\infty} a_n e^{2\pi i n t}.$$

If two Fourier series are equal, then they are equal term by term, so we get

$$b_n = \frac{a_n}{1 - e^{-2\pi i n p}}. \tag{2.4}$$

We have solved the first part of our problem; we have found the Fourier coefficients of the unknown function g. Now we have to sum the series, and it is here that we run into trouble. Each coefficient b_n contains the troublesome term $(1 - e^{-2\pi i n p})$ in the denominator, and this number is very small for some values of n; in fact, it is 0 for some values of n if p is rational.

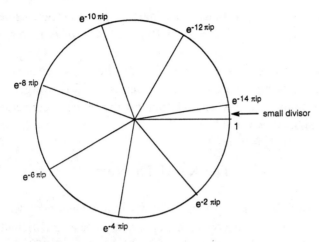

Figure 5. The problem of small divisors—and of denominators with the value zero—can be illustrated by a circle of radius 1 in the complex plane. The point $e^{-2\pi ip}$ is the point with coordinates $(\cos 2\pi p, -\sin 2\pi p)$. If p is a rational number $\frac{a}{b}$, and n is a multiple of b, then np is a whole number, $e^{-2\pi inp} = 1$, and formula (2.4) for the coefficients of our Fourier series contains a zero in the denominator. If p is irrational, the number np is never a whole number of turns, but it can come arbitrarily close to an integer, and the denominator $1 - e^{-2\pi inp}$ is then very small. The problem of knowing whether a series with small divisors converges or diverges is extraordinarily difficult.

To see this, consider the circle of radius 1 in the complex plane shown in Figure 5; the point $(1,0)$ is thus the complex number 1. The point $e^{-2\pi ip}$ is the point with coordinates $(\cos 2\pi p, -\sin 2\pi p)$. If p is a rational number $\frac{a}{b}$, then np is a whole number when n is a multiple of the denominator b. In that case, $e^{-2\pi inp} = 1$, and the problem has no solution: our formula (2.4) contains a zero in the denominator. (Remember that $e^{i\theta} = \cos\theta + i\sin\theta$, so $e^{i\theta} = 1$ when $\theta = 2\pi$. If we multiply 2π by an integer k, we make k revolutions of the circle and land again on the complex number 1.)

What happens when p is irrational? In this case coefficients b_n in equation (2.4) exist, but we still need to watch out. If p is irrational, the number np is never a whole number of turns, but it can come arbitrarily close to an integer, and the denominator $1 - e^{-2\pi inp}$ is then very small. For example, the astronomical year does not contain exactly 365 days, it contains 365.24 ... This will give us small divisors (and big coefficients) roughly every 365th coefficient, and much bigger coefficients yet for all multiples of (365 x 4) + 1 = 1461 (that is, 1461 days is very close to four years).

The small divisors $1-e^{-2\pi i n p}$ will divide the a_n, which are themselves small when n is big, since the Fourier series for f converges. But will these a_n become small fast enough to allow the series to converge, or will the small divisors win out?

The KAM Theorem

In the case of the solar system, it seems that Weierstrass was inclined to believe that the series converges, apparently because Dirichlet, known for his mathematical rigor, said before his death in 1859 that he had found a way to approximate solutions to the n-body problem [Moser, p. 8]. The American George Birkhoff and Frenchman Henri Poincaré believed the series did not converge; Poincaré was awarded a prize by the king of Sweden in 1885 for a paper that, in addition to outlining most modern methods now used in dynamical systems, seemed to indicate that in practice no orbit in a system of n bodies would remain stable forever (that such orbits are rare to the extent of having zero probability.)

The first indications that this conclusion was wrong came in 1954, when Kolmogorov, speaking in Stockholm at the International Congress of Mathematicians, sketched the general lines of a proof that orbits (of planets, of particles,...) can be stable when there is *a priori* no reason for them to be so (i.e., no "law of conservation" enforcing that stability.)[1] Proofs of the theorem were published in 1962 and 1963 by Vladimir Arnold and Jürgen Moser.

The theorem, now called the KAM theorem, applies to all nondissipative systems of classical mechanics (systems without friction). It shows

---

[1]If planets did not attract each other, then the solar system would be a linear system, and the amount of kinetic energy would remain constant, as would the amount of angular momentum. These "conservation laws" would force the solutions to the system of linear equations describing the motions of the planets to remain on a torus. But planets do attract each other, trading kinetic energy and angular momentum back and forth. In such a nonlinear system, one would expect solutions to wander all over the energy hypersurface.

that if one thinks of such systems as a struggle between order and disorder, order is more powerful than had been thought: under some conditions, these systems are inherently stable. The difference between stability and chaos can hinge on a delicate question in number theory: to what extent irrational numbers can be approximated by rational numbers. (For a mathematical discussion of KAM, see [Arnold, Avez], [Hubbard], and [Moser]; for a popularization, see [Hubbard, Hubbard].)

KAM does not exhaust the subject of small divisors; in August 1994, Jean-Christophe Yoccoz received the Fields Medal for finding which Fourier series in a related problem converge.

Computing Fourier Coefficients with Integrals

To compute the Fourier coefficients of a periodic function f of period 1, we multiply f by the functions $\sin 2\pi kx$ and $\cos 2\pi kx$ (sines and cosines of integer frequencies k). Since the functions $\sin 2\pi kx$ and $\cos 2\pi kx$ oscillate between $+1$ and -1, this multiplication produces a function whose graph oscillates between the graphs of $+f$ and $-f$.

The integral of this product (the area delimited by the graph of the new function) is the Fourier coefficient at a given frequency. The negative area (below the x-axis) is subtracted from the positive area (above the x-axis). We integrate over the period of the function (here, between 0 and 1; for a function periodic of periodic 2π, between 0 and 2π). Figures 2–5 illustrate this procedure for the function shown in Figure 1.

At very high frequencies, the Fourier coefficients of a smooth function tend towards zero. The function changes slowly relative to the rapid, high-frequency oscillations, and these oscillations tend to delimit almost equal amounts of negative and positive areas, producing very small coefficients. But it's not true that Fourier coefficients always get smaller as the frequency increases. For the function shown in Figure 1, the coefficient for cosine of frequency 100, is small (see Figure 4), but it is larger than that for frequency 7, which is zero (see Figure 5). (For this function, all the coefficients for $\cos nx$ are zero, when n is odd.)

137

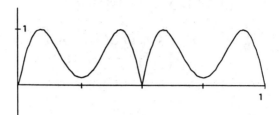

Figure 1. The function $f(x) = |\sin(3\sin 2\pi x)|$

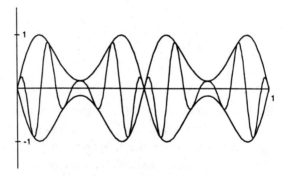

Figure 2. The function $f(x)\cos 8(2\pi x)$. Its integral—the Fourier coefficient of f for the cosine of frequency 8—is -0.2311.

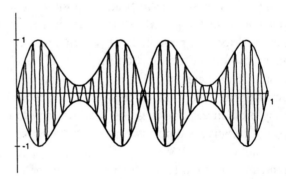

Figure 3. The Fourier coefficient of f for the cosine of frequency 30 is -1.327×10^{-2}.

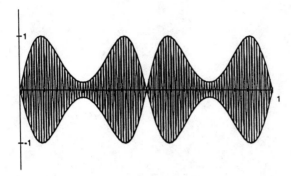

Figure 4. The Fourier coefficient of f for the cosine of frequency 100 is -1.192×10^{-3}.

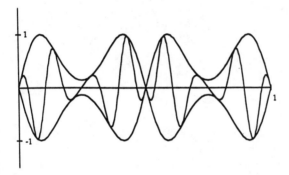

Figure 5. The Fourier coefficient of f for the cosine of frequency 7 is 0.

The Fast Fourier Transform

An algorithm is a recipe for doing computations. When children multiply two-digit numbers or "borrow" to subtract one number from another, they are using algorithms. More sophisticated algorithms enable computers to carry out computations that would otherwise be endless. The modern algorithm that has most transformed our society is the FFT (fast Fourier transform). "Whole industries are changed from slow to fast by this one idea—which is pure mathematics," writes mathematician Gilbert Strang of the Massachusetts Institute of Technology [Strang, p. 290].

The FFT cuts from n^2 to $n \log n$ the number of computations necessary to compute the Fourier transform of a signal with n values. (The logarithm base b of n, written $\log_b n$, is the power to which one must raise the base b to obtain n: $\log_2 4 = 2$, since $2^2 = 4$; $\log_2 8 = 3$, since $2^3 = 8$; $\log_{10} 100 = 2$, since $10^2 = 100$. In other words, $\log_b n$ is roughly the number of digits of n written in base b: $\log_{10} 1\,000 = 3$; $\log_{10} 374\,113 \approx 5.57$; $\log_{10} 1\,000\,000 = 6$.)

The bigger n is, the more impressive the gain in speed. If $n = 2^{10} = 1024$, then $n^2 = 1\,048\,576$, while $n \log_2 n = (1024)(10) = 10\,240$, a difference of a factor of about 100. If $n = 2^{20} = 1\,048\,576$, then $n^2 = 1\,099\,511\,627\,776$, but $n \log_2 n = 20\,971\,520$: a difference of a factor of about 50 000. With the FFT and a good computer, one can compute π to a billion digits in less than an hour; with the same computer but without the FFT, the job would take almost ten thousand years.

The idea underlying the FFT was discovered by Carl Friedrich Gauss, probably in 1805, two years before Fourier presented his memoir to the Academy of Sciences in Paris; it was published only after his death.

(See [Heideman et al.] for a historical discussion.) The algorithm was rediscovered and used as a computer program by James Cooley and John Tukey in 1965.

"Like the fast wavelet transform, and most efficient algorithms, it is a divide and conquer algorithm," says Martin Vetterli of the University of California at Berkeley. One way to understand it (described in [Strang]) is as a clever factorization of a particular matrix; for those who learned in school to manipulate matrices without understanding why, the FFT is one possible answer. (Some mathematicians and engineers prefer to describe the FFT as splitting a single sum into double sums; for this more traditional approach, see for example [Duhamel, Vetterli].)

The Slow Fourier Transform

We will begin by computing a few Fourier coefficients without the FFT, using complex numbers to simplify matters. We will call c_k the Fourier coefficient at frequency k of the signal $f(x)$: the number that tells how much of frequency k is present in the signal. (In *The Fourier Transform* we used k to represent frequency in Fourier series, and ξ or τ in Fourier transforms, explaining that by convention k represents integer variables, and ξ and τ represent continuous variables. We use k here because the fast Fourier transform is a discrete form of the Fourier transform.) We will speak of a signal $f(x)$ that we know for $0 \leq x \leq 1$. Then the formula for computing the coefficients is:

$$c_k = \int_0^1 f(x)e^{2\pi i k x} \, dx. \tag{4.1}$$

Integrating isn't simple. Formulas giving exact values exist in some cases; more often one must make do with approximations. To get these approximations, we sample the signal at regularly spaced points and obtain average values for each frequency (multiplying the sampled values by the exponential of the frequency in question, adding the products, and dividing the total by the number of samples). We restrict ourselves—arbitrarily, to make the algorithm work better—to a number of frequencies equal to the number of samples.

Rephrasing this in mathematical language: we divide the period of the signal by 2^N and measure the signal at points $\frac{j}{2^N}$ for $j = 0, 1, \ldots,$ $2^N - 1$. Then we multiply the exponential of each frequency k by the value of the signal at each point $\frac{j}{2^N}$. (Since the number of frequencies used is equal to the number of points, k also goes from 0 to $2^N - 1$.) For each frequency, all these products are added, and the result is divided by 2^N. In other words:

$$c_k = \frac{1}{2^N} \sum_{j=0}^{2^N - 1} f\left(\frac{j}{2^N}\right) e^{2\pi i k \frac{j}{2^N}}, k = 0, 1, \ldots, 2^N - 1. \qquad (4.2)$$

Let's take a simple example, where $N = 2$. Then we need to compute four coefficients c_k (for $k = 0, 1, 2, 3$). We compute them using four samples of the signal: at 0, $\frac{1}{4}$, $\frac{1}{2}$ and $\frac{3}{4}$ (the values of $\frac{j}{2^N}$, for $j = 0, 1, 2, 3$). Let's take as our signal the function $f(x) = x^2$; then $f\left(\frac{j}{2^N}\right)$ becomes $\left(\frac{j}{4}\right)^2$, and we have, for each frequency k, four terms to sum:

1. for $j = 0$, $(0)^2 \, e^{2\pi i k \frac{0}{4}} = 0$
2. for $j = 1$, $\left(\frac{1}{4}\right)^2 e^{2\pi i k \frac{1}{4}} = \frac{1}{16} e^{\pi i k \frac{1}{2}}$
3. for $j = 2$, $\left(\frac{2}{4}\right)^2 e^{2\pi i k \frac{2}{4}} = \frac{1}{4} e^{\pi i k}$
4. for $j = 3$, $\left(\frac{3}{4}\right)^2 e^{2\pi i k \frac{3}{4}} = \frac{9}{16} e^{3\pi i k \frac{1}{2}}$

Now we can actually compute the coefficients. The frequency $k = 0$ gives exponents with the value zero, and any number with an exponent of zero equals 1. So for the first coefficient c_0 we have:

$$c_0 = \frac{1}{4}\left(0 + \frac{1}{16} + \frac{1}{4} + \frac{9}{16}\right) = \frac{14}{64}$$

The coefficient for $k = 1$ is a little more complicated. We use the formula $e^{\pi i/2} = i$ (which follows directly from the formula $e^{i\theta} = \cos\theta + i\sin\theta$, discussed in *The Fourier Transform*, p. 117). We get:

$$c_1 = \frac{1}{4}\left(0 + \frac{i}{16} - \frac{1}{4} - \frac{9i}{16}\right) = \frac{1}{4}\left(-\frac{1}{4} - \frac{i}{2}\right) = -\frac{1}{8}\left(i + \frac{1}{2}\right).$$

The other coefficients are computed the same way.

Computing Fourier coefficients this way is straightforward but tedious. Measuring four frequencies at four points is hardly sufficient; a more realistic number is $2^{10} = 1024$. For each of 1024 frequencies, one must add the products of 1024 multiplications: more than a million computations for a single transform. One could easily get discouraged.

A Short-Cut Via Matrices

Let's see if we can do better using matrices. A matrix is a rectangular arrangement of numbers; multiplying matrices is easy but not intuitive. Two matrices, A and B, can be multiplied if the number of columns in A equals the number of rows in B. (Order matters: it may be possible to multiply AB but not BA. Even if both matrices have the same number of rows and columns, AB will usually not equal BA.)

In the example below, the first element of the product AB (the matrix at bottom right) is obtained by multiplying, one by one, the elements of the first *row* of A (on the left) by the elements of the first *column* of B (top right) and adding them together: $(-1 \times 1) + (1 \times -1) + (0 \times 1)$. The total (called the *scalar product* or *dot product*) equals -2.

$$
\begin{bmatrix} & & \\ & B & \\ & & \end{bmatrix}
\qquad
\begin{bmatrix} 1 & 0 & 1 \\ -1 & 2 & 1 \\ 1 & 1 & 0 \end{bmatrix}
$$

$$
\begin{bmatrix} & & \\ & A & \\ & & \end{bmatrix}
\begin{bmatrix} & & \\ & AB & \\ & & \end{bmatrix}
\begin{bmatrix} -1 & 1 & 0 \\ 2 & 3 & 2 \end{bmatrix}
\begin{bmatrix} -2 & 2 & 0 \\ 1 & 8 & 5 \end{bmatrix}
$$

Next one goes to the first row of A and the second column of B: $(-1 \times 0) + (1 \times 2) + (0 \times 1) = 2 \dots$ (This way of arranging matrices, with B above AB, is not standard. But especially when matrices are big, it makes computations less confusing: the scalar product of the ith row of A and the jth column of B lies at the intersection of that row and column.)

Now we can rewrite equation (4.2) in terms of matrices; see Figure 1. In the A matrix (which we will call F_{2N} for Fourier) we put the numbers $e^{2\pi i k \frac{j}{2N}}$, letting j vary horizontally and k vertically. The B matrix (called

$$\begin{bmatrix} f(\frac{0}{2^N}) \\ f(\frac{1}{2^N}) \\ \cdots \\ f(\frac{2^N-1}{2^N}) \end{bmatrix}$$

$$\begin{bmatrix} e^{2\pi i 0 \frac{0}{2^N}} & e^{2\pi i 0 \frac{1}{2^N}} & \cdots & e^{2\pi i 0 \frac{2^N-1}{2^N}} \\ e^{2\pi i 1 \frac{0}{2^N}} & e^{2\pi i 1 \frac{1}{2^N}} & \cdots & e^{2\pi i 1 \frac{2^N-1}{2^N}} \\ \cdots & \cdots & \cdots & \cdots \\ e^{2\pi i (2^N-1) \frac{0}{2^N}} & \cdots & \cdots & e^{2\pi i \frac{(2^N-1)(2^N-1)}{2^N}} \end{bmatrix} \begin{bmatrix} 2^N c_0 \\ 2^N c_1 \\ \cdots \\ 2^N c_{2^N-1} \end{bmatrix}$$

Figure 1. The computation of Fourier coefficients is represented here with matrices. The matrix at upper right contains the sampled values of the signal, while the matrix to the left contains the numbers $e^{2\pi i k \frac{j}{2^N}}$. The matrix at lower right—the product of the other two matrices—contains the Fourier coefficients, multiplied by $2N$.

a vector since it has only one column) contains the sampled values of the signal at $\frac{j}{2^N}$ (for $j = 0, 1, \ldots, 2^N - 1$). Dividing the product of the two matrices by 2^N gives the coefficients $c_0, c_1, \ldots, c_{2^N-1}$.

A Clever Factorization

Unfortunately, this exercise in itself doesn't get us anywhere. Our matrix F_{2^N} has 2^{2N} elements (2^{20} for the fairly realistic case where $N = 10$), and we still need the same 2^{20} multiplications—more than a million—to compute the coefficients. The trick of the FFT is to factor the matrix F_{2^N} into three other matrices.

The underlying idea is to rearrange the numbers that have to be multiplied together so as to avoid doing the same computation twice. A schoolchild told to multiply by hand $9\,996\,496 \times 8\,426\,735$ can make the task easier by putting the second number on top of the first and multiplying $8\,426\,735$ by 9 only once, copying the product where appropriate. When we compute a Fourier transform by multiplying the kth frequency by the jth sample, we will land on more than one way of getting the same product kj. If $kj = 24$, for example, then we have

$$kj = (1 \times 24) = (24 \times 1) = (2 \times 12) = (12 \times 2) = (3 \times 8) = \cdots .$$

Gauss figured out a way to take advantage of such coincidences while trying to determine orbits of asteroids. Assuming an orbit can be expressed as a trigonometric polynomial, and knowing the asteroid's position at certain times, he wanted to determine the intermediate positions by interpolation. (One can think of the orbit as a sampled signal. From the sampled values, Gauss determined the coefficients of the polynomial—computing its Fourier transform—and then he evaluated the polynomial—inverting the transform. Of course he didn't use that terminology.) The algorithm he developed in the process "greatly reduces the tediousness of mechanical calculations," he wrote in a paper he himself never published [Gauss, p. 307]. (Gauss's paper is in Latin; for an English account, see [Goldstine].)

Gauss did not write his algorithm as a factorization of matrices (matrices first appeared in a published work, by Arthur Cayley, in 1858 [Cayley]). But in matrix form his algorithm is written:

$$[F_{2N}] = \underbrace{\begin{bmatrix} I_{2N-1} & D_{2N-1} \\ I_{2N-1} & -D_{2N-1} \end{bmatrix}}_{[1]} \underbrace{\begin{bmatrix} F_{2N-1} & 0 \\ 0 & F_{2N-1} \end{bmatrix}}_{[2]} \underbrace{\begin{bmatrix} \text{shuffle} \end{bmatrix}}_{[3]} .$$

The first of these new matrices, labeled [1], contains four sub-matrices (called I, D, I and $-D$). All four are very sparse: all the elements are 0 except those on the main diagonals. In the sub-matrices called I, for "identity," the elements on the main diagonals all have the value 1. For example, when $N = 3$ and $\omega = e^{2\pi i/2^N}$:

$$I_{2N-1} = \begin{bmatrix} 1 & 0 & 0 & 0 \\ 0 & 1 & 0 & 0 \\ 0 & 0 & 1 & 0 \\ 0 & 0 & 0 & 1 \end{bmatrix} \text{ and } D_{2N-1} = \begin{bmatrix} \omega^0 & 0 & 0 & 0 \\ 0 & \omega^1 & 0 & 0 \\ 0 & 0 & \omega^2 & 0 \\ 0 & 0 & 0 & \omega^3 \end{bmatrix} .$$

$$
\begin{bmatrix}
1 & 0 & 0 & 0 & 0 & 0 & 0 & 0 \\
0 & 0 & 1 & 0 & 0 & 0 & 0 & 0 \\
0 & 0 & 0 & 0 & 1 & 0 & 0 & 0 \\
0 & 0 & 0 & 0 & 0 & 0 & 1 & 0 \\
0 & 1 & 0 & 0 & 0 & 0 & 0 & 0 \\
0 & 0 & 0 & 1 & 0 & 0 & 0 & 0 \\
0 & 0 & 0 & 0 & 0 & 1 & 0 & 0 \\
0 & 0 & 0 & 0 & 0 & 0 & 0 & 1
\end{bmatrix}
\begin{bmatrix}
s_0 \\ s_1 \\ s_2 \\ s_3 \\ s_4 \\ s_5 \\ s_6 \\ s_7
\end{bmatrix}
=
\begin{bmatrix}
s_0 \\ s_2 \\ s_4 \\ s_6 \\ s_1 \\ s_3 \\ s_5 \\ s_7
\end{bmatrix}
$$

Figure 2. An example of a shuffle. The first matrix, the shuffle matrix, "unshuffles" the entries s_0 to s_7 of the second matrix, putting the even entries (s_0, s_2, etc.) on top and the odd entries (s_1, s_3, etc.) on the bottom.

The second [2] of these new matrices is half empty, with half as many nonzero elements as the original matrix F_{2N}. The third new matrix [3] is the shuffle. As shown in Figure 2, a shuffle matrix shuffles the elements of the vector (i.e., the sampled values of the signal) without changing their values, like shuffling a deck of cards. (Actually, it "unshuffles" them: it takes every other "card"—the even entries—and puts it in the top half of the "deck," putting the odd entries in the bottom half.) Figure 2 shows an 8 x 8 shuffle matrix shuffling 8 samples, s_0 to s_7. (Here we use the traditional format for matrix multiplication.)

This factorization cuts the work of computing a Fourier transform almost in half. Instead of a million computations, we now have about half a million: 2 x (512 x 512) for the two sub-matrices in [2] and about 1,000 for the matrix [1]. (Often the shuffling isn't counted, being considered part of administrative overhead. This is perhaps unwise, considering that administrative overhead charged by research universities on government grants typically runs some 40 percent. But in any case, when we say that we need $n \log n$ computations to compute a Fourier transform with the FFT, we really mean $cn \log n$, the constant c depending on the details of implementation.)

But why stop there? The two sub-matrices in matrix [2] can themselves be factorized the same way, giving new sub-matrices with $2^{N-2} =$
$2^8 = 256$ elements each, which themselves can be factorized ... The
number of computations falls precipitously in 10 (since $N = 10$) steps:

1. From 1 million to $\frac{1}{2}$ million $+ \approx 1,000$

2. ... to $\frac{1}{4}$ million $+ \approx 1,000 + \approx 1,000 = \frac{1}{4}$ million $+ \approx 2,000$

...

10. ... to $\frac{1}{1024}$ million $+ \approx 10,000 \approx 11,000$.

If one had to do the work by hand, this would still be excessive, but
with the advent of computers the FFT became a powerful tool.

The Continuous Wavelet Transform

In the continuous wavelet transform, a function ψ ("psi"), which in practice looks like a little wave, is used to create a family of wavelets $\psi(at+b)$ where a and b are real numbers, a dilating (compressing or stretching) the function ψ and b translating (displacing) it. The word *continuous* refers to the transform, not the wavelets, although people sometimes speak of "continuous wavelets."

The continuous wavelet transform turns a signal $f(t)$ into a function with two variables (scale and time), which one can call $c(a, b)$:

$$c(a, b) = \int f(t)\, \psi(at + b)\, dt.$$

This transformation is in theory infinitely redundant, but it can be useful in recognizing certain characteristics of a signal. In addition, the extreme redundancy is less of a problem than one might imagine; a number of researchers have found ways of rapidly extracting the essential information from these redundant transforms.

One such method reduces a redundant transform to its *skeleton*. When certain signals are represented by a continuous wavelet transform, all the significant information of the signal is contained in curves, or "ridges," says Bruno Torrésani of the French Centre National de Recherche Scientifique, who works at the University of Aix-Marseille II. These are essentially the points in the time-frequency plane "where the natural frequency of the translated and dilated wavelet coincides with the local frequency, or one of the local frequencies, of the signal." These ridges form the skeleton of the transform.

Torrésani, working with Richard Kronland-Martinet in Marseille and Bernard Escudié in Lyon, found algorithms that exploit the redundancy of a continuous transform in order to calculate the skeleton rapidly (see [Tchamitchian, Torrésani], [Delprat et al.]). Others in Marseille have worked on the method, including Nathalie Delprat, Philippe Guillemain, and Philippe Tchamitchian, who have applied it to music, and Caroline Gonnet, who has worked with Torrésani on the skeletons of images [Gonnet, Torrésani]. In a French-Italian project (VIRGO), Jean-Michel Innocent is trying to apply it to the detection of gravitational waves emitted, for example, by a collapsing binary star. Such waves are predicted by the theory of general relativity but have never been observed. "The big problem is to separate the signal from background noise, which is enormous in this case," says Torrésani, who is also working on the problem of skeletons in the presence of substantial noise.

The skeleton method applies to signals with a locally narrow bandwidth—some speech signals, for example. That is, signals where one can associate a well-defined frequency (or frequencies) to each instant of the signal. This is not the case for signals with singularities—places where the signal varies rapidly, such as at the edges of images. For these signals, Stéphane Mallat and Wen Liang Hwang of the Courant Institute of Mathematical Sciences [Mallat, Hwang] found a different way to reduce a redundant transform to something more manageable: calculating the transform's maximum wavelet coefficients, called *wavelet maxima* (see p. 66). Ways to reconstruct a signal from its wavelet maxima were developed by Mallat and Sifen Zhong [Mallat, Zhong].

Discrete Wavelet Transforms

In a discrete wavelet transform, a wavelet is translated and dilated only by discrete values. Most often dilation is by a power of 2 (sometimes called *dyadic*). That is, one uses wavelets only of the form

$$\psi(2^k t + l), \quad \text{with } k \text{ and } l \text{ whole numbers.}$$

Orthogonal wavelets (see *Orthogonality and Scalar Products*, p. 153 and *Multiresolution*, p. 165) are a special case of discrete wavelets. They give a representation without redundancy and lend themselves to fast algorithms. It is much less evident how to construct them; until Mallat and Yves Meyer developed their multiresolution theory, finding a single orthogonal wavelet basis was a major achievement.

Orthogonality and Scalar Products

Those unfamiliar with the concepts involved in Fourier analysis and wavelets may have found some statements in the main article a bit mysterious. We said that the Fourier transform breaks down a signal by frequency, and that the wavelet transform breaks down a signal into components of different scales by comparing the signal to wavelets of different sizes. In both cases, we said, this is done by integration: multiplying the signal by the analyzing function (sines and cosines or wavelets) and integrating the product. Why does calculating integrals allow us to decompose a signal? If you glance at any signal-processing or wavelet text, you'll see that mathematicians and engineers refer to computing Fourier or wavelet coefficients as taking the *scalar product* of the signal and the analyzing function. (Sometimes, to add to the confusion, they call it a *dot product* or *inner product*.) If the technique used is integration, why drag in dot, scalar, or inner products?

We've also talked about orthogonal wavelets, historically far harder to create than the wavelets used in a continuous transform, and offering the advantages (and disadvantages) of conciseness, while allowing for perfect reconstruction of the original signal. Your dictionary will tell you that "orthogonal" means "perpendicular." What does it mean for wavelets to be perpendicular? What do right angles have to do with encoding signals?

The purpose of this section is to answer those questions and in doing so to better understand why people are interested in orthogonal transforms. Conciseness is not the only objective; in fact, some discrete but non-orthogonal transforms are equally concise. But the geometrical properties of orthonormal wavelets makes the transform easy to compute: each coefficient is computed with a single scalar product, and that

153

computation is independent of the computation of the other coefficients of the transform.

To see this, we will use geometrical ideas that are elementary in two or three dimensions: adding vectors and calculating scalar products, often taught in high school. These same ideas are much less intuitive and acquire much greater power when applied to infinite-dimensional function spaces. We need, then, to think of a function or signal as a single point in an infinite-dimensional space. If you find this strange, you are in good company; to mathematicians it is now second nature, but previous generations of mathematicians struggled with the concept. Learning to think of functions this way—and thus to think of functions sharing certain characteristics as elements of the same *function space*, rather than thinking of each function as an individual defined by a formula—is an important milestone in the coming of age of a mathematician.

A Function as a Point in an Infinite-Dimensional Space

A point on a line is determined by a single number, a point in the plane by two (its two coordinates), and a point in three-dimensional space by its three coordinates. To define a function one must know all its values: an infinity of numbers. A function, such as a wavelet or a signal, can be thought of as a single point in an infinite-dimensional space.

We can represent a point in an n-dimensional space by its *projections* onto n axes. This is easy to visualize in two or three dimensions. In school, the axes are chosen to be perpendicular (orthogonal), so that if you draw a line from the point (3,5) perpendicular to the x-axis it will intersect the x-axis at the point 3; we say that the projection of the point on the x-axis is 3. (Similarly, its projection on the y-axis is 5.)

By analogy, we will use the geometric vocabulary to talk about functions as points in an infinite-dimensional space. Given a signal, we may wish to decompose it with Fourier analysis, wavelets, or some other basis[1] depending on the nature of the signal (music, picture...) or the job to be done (analysis, compression...).

---

[1] A *basis* is a family of analyzing functions—a "mother" wavelet and its dilates and translates, for example—that can represent any signal. There are infinitely many different orthogonal wavelet bases, each created by a different mother wavelet.

Each basis function determines the direction of an axis, so choosing a basis means choosing a system of axes. Decomposing a signal means representing it by its infinitely many coordinates, by projecting it onto the infinitely many axes of the basis. The coefficients (wavelet coefficients, Fourier coefficients, and so on) simply measure where on each axis the projection lands. The basis is *orthogonal* if all its functions—thus the axes on which the signal is projected—are perpendicular to each other.

This way of speaking assumes that the words *length* and *angle* have a meaning in higher dimensions, which isn't obvious. To make this geometrical language more natural, we will identify a *point* with the *vector* that goes from the origin to the point. A vector and a point are the same thing, but it is more natural to talk of the angle between two vectors or the length of a vector, than to talk of the angle between two points or the length of a point. But even with this change of terminology, what meaning can we attach to the length of a vector in 17 dimensions, or the angle between two vectors in infinite dimensions?

It is the scalar product that will enable us to construct a geometry for these spaces.

Scalar Products

The scalar product $< \vec{a}, \vec{b} >$ of two vectors $\vec{a}$ and $\vec{b}$, each with n coordinates, is:

$$\left\langle \vec{a}, \vec{b} \right\rangle = \left\langle \begin{bmatrix} a_1 \\ a_2 \\ \vdots \\ a_n \end{bmatrix}, \begin{bmatrix} b_1 \\ b_2 \\ \vdots \\ b_n \end{bmatrix} \right\rangle = a_1 b_1 + a_2 b_2 + \ldots + a_n b_n. \qquad (6.1)$$

If $n = 3$, $\vec{a}$ is the vector (3, 6, 1), and $\vec{b}$ the vector (2, 5, 3), then $< \vec{a}, \vec{b} >= 6 + 30 + 3 = 39$. Note that a scalar product is a *number*, not a vector.[2]

---

[2] Actually, the scalar product we describe here is a particular scalar product, known as the *standard inner product* or the *dot product*, the latter because it is sometimes denoted with a dot: $\vec{a} \cdot \vec{b}$. In the context of abstract vector spaces, other scalar products exist.

We can easily see how the definition (6.1) of a scalar product applies in higher dimensions—the columns become longer and the arithmetic more tedious, but computing a scalar product in 10 dimensions, or in 100, is straightforward.

Thinking about lengths and angles in higher dimensions is harder, yet we've talked nonchalantly of infinitely many basis functions all perpendicular to each other. Definition (6.1), along with the following two equations, enables us to think geometrically in higher dimensions. Equations (6.2) and (6.3) are theorems in two and three dimensions and definitions in higher dimensions. Equation (6.2) tells us that the length of a vector is the square root of the scalar product of the vector with itself:

$$|\vec{a}|^2 = <\vec{a}, \vec{a}>, \tag{6.2}$$

where $|\vec{a}|$ is the length of $\vec{a}$. (In two and three dimensions, this is a version of the Pythagorean theorem: the vector $\vec{a}$ with coordinates (x, y) is the hypotenuse of a right triangle of which the other two sides have lengths x and y. By formula (6.1), $<\vec{a}, \vec{a}> = x^2 + y^2$.)

Equation (6.3) defines the angle θ between two vectors $\vec{a}$ and $\vec{b}$:

$$<\vec{a}, \vec{b}> = |\vec{a}| \, |\vec{b}| \cos\theta. \tag{6.3}$$

(Since $\cos\theta \leq 1$, this leads directly to the Schwarz inequality: $|\vec{v}|^2 |\vec{w}|^2 \geq |<\vec{v}, \vec{w}>|^2$, central to proving the Heisenberg uncertainty principle, as we see on p. 281. And since $\cos 90° = 0$, we see that two orthogonal vectors have a scalar product of zero.)

The notions of length and angle, which are perfectly clear in two and three dimensions, thus acquire definitions in higher dimensions. We still can't visualize space in 17 dimensions, or in infinite dimensions, but we can work by analogy. And definition (6.1), which extends so easily to higher dimensions, can make it easier to accept the rather bold idea that these familiar notions mean something in unimaginable infinite dimensional spaces.

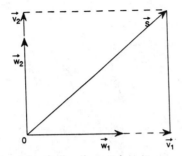

Figure 1. A signal (the vector $\vec{s}$) decomposed in an orthonormal basis in two dimensions, with the elements $\vec{w}_1$ and $\vec{w}_2$. Projecting the signal onto axes pointing in the direction of the basis elements gives two new vectors, $\vec{v}_1$ and $\vec{v}_2$. The signal equals their sum: $\vec{s} = \vec{v}_1 + \vec{v}_2$.

Calculating Coefficients with Scalar Products

Our goal is to show that a coefficient of an *orthonormal* transform can be computed with a single scalar product of two vectors, the signal and an element of the basis. (*Orthonormal* means that all the elements of an orthogonal basis are *normalized* to have length 1.) The computation is relatively simple, since it is done independently of all the other basis functions, and each coefficient encodes information that is encoded nowhere else.

Consider a simple example. In Figure 1 we decompose a signal (the vector $\vec{s}$) in an orthonormal basis in two dimensions. The basis has two elements: $\vec{w}_1$ and $\vec{w}_2$. (These vectors could be wavelets, but they could just as well be elements of some other basis having nothing to do with wavelets.[3]) We project the signal onto axes that point in the direction of the basis elements, creating two new vectors $\vec{v}_1$ and $\vec{v}_2$ (starting also at 0). Then $\vec{s} = \vec{v}_1 + \vec{v}_2$.

(Recall that we add two vectors in the plane by adding their coordinates. The sum of the two x-coordinates gives the x-coordinate of the new vector, and the sum of the y-coordinates gives the y-coordinate of the new vector, giving a new vector that is the diagonal of the parallelogram

---

[3]A proof that the trigonometric functions form an orthonormal basis is given in the Appendix, p. 291.

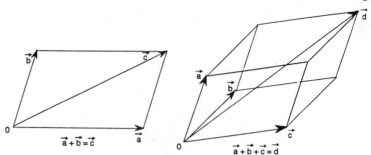

Figure 2. Vectors are added by adding their coordinates, producing a new vector. In two dimensions the new vector is the diagonal of the parallelogram of which the two original vectors are adjacent sides. In three dimensions the new vector is the diagonal of the parallelepiped spanned by the three vectors.

of which the original two vectors are two adjacent sides: $\vec{a} + \vec{b} = \vec{c}$, as shown in Figure 2. The analogous relationship holds in three or more dimensions.)

Computing our coefficients c_1 and c_2 amounts to expressing the vectors $\vec{v}_1$ and $\vec{v}_2$ in terms of our basis elements:

$$\vec{s} = \vec{v}_1 + \vec{v}_2 = c_1\vec{w}_1 + c_2\vec{w}_2. \tag{6.4}$$

It is now straightforward to show that a coefficient in an orthonormal transform is the scalar product of the signal and an element of the basis: $c_1 = <\vec{w}_1, \vec{s}>$. We take the scalar product of each side of equation (6.4) with $\vec{w}_1$:

$$<\vec{s}, \vec{w}_1> = <(c_1\vec{w}_1 + c_2\vec{w}_2), \vec{w}_1> = c_1 <\vec{w}_1, \vec{w}_1> + c_2 <\vec{w}_2, \vec{w}_1> .$$

The basis is orthonormal, so $\vec{w}_1$ and $\vec{w}_2$ have length 1; using Equation (6.2), we have $<\vec{w}_1, \vec{w}_1> = 1$. Two orthogonal vectors have the scalar product 0, so $<\vec{w}_2, \vec{w}_1> = 0$. Therefore,

$$<\vec{s}, \vec{w}_1> = c_1.$$

Our basis has only two elements, but if it had an infinite number of elements the scalar product of w_1 with each of the others would still be 0, and the scalar product of w_1 with itself would still be 1, leaving us with

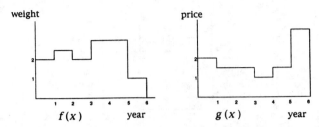

Figure 3. Left: the function f representing steel production over a period of 6 years. Right: the function g representing the price of steel. The integral of the product of these functions gives the total value of the steel produced, shown in Figure 4. The same information can be given by a scalar product: in this case, the integral and the scalar product coincide exactly.

$< \vec{s}, \vec{w}_1 > = c_1$. So while we have, for simplicity, posed our problem in two dimensions, we are actually talking about the real-life situation of a signal in an n-dimensional space.

And the Integrals?

We've shown that a coefficient in an orthonormal transform can be computed with a single scalar product. Earlier we talked about computing wavelet coefficients with integrals. Both are right, because a scalar product and an integral are essentially the same thing.

Let's consider a case where the two coincide exactly: the functions $f(t)$ that are defined for $0 \le t \le n$, and are constant, except possibly at the integers. Such bar graphs are often used, for example, in representing annual industrial production. Let's imagine a function f that represents the annual production of steel, and another, g, that represents its average price by weight from one year to the next, as shown in Figure 3.

(For obvious reasons we remain vague about the units of weight and of price.) Then the integral

$$\int_0^6 f(t)g(t)\,dt$$

gives the total value of the production of steel over 6 years:

The same information can also be given by a scalar product:

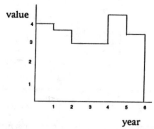

year

Figure 4. The integral of the product fg gives the total value of steel produced over 6 years.

$$\langle \vec{f}, \vec{g} \rangle = \left\langle \begin{bmatrix} 2 \\ 2.5 \\ 2 \\ 3 \\ 3 \\ 1 \end{bmatrix}, \begin{bmatrix} 2 \\ 1.5 \\ 1.5 \\ 1 \\ 1.5 \\ 3.5 \end{bmatrix} \right\rangle = 4+3.75+3+3+4.5+3.5 = 21.75.$$

In this case the integral and the scalar product are identical. Approximating an integral by a Riemann sum is equivalent to approximating the function being integrated by a bar graph; so computing the integral of the product of two functions using a Riemann sum comes down to computing the scalar product of the functions. The result may be interpreted geometrically as "space under a curve," but for the computation itself the geometry recedes to the background as one carries out a series of multiplications and additions.

This example makes it reasonable to define the scalar product on the space of functions on $[a, b]$ (more precisely, the space $L^2[a, b]$ of square integrable functions) to be

$$< f, g >= \int_a^b f(t)g(t)\, dt.$$

(This concordance between scalar product and integral makes it clear that the notion of the length of a function—a wavelet, for example—is not intuitive. Its length is the square root of the integral of its square. So to keep the "length" constant, skinny dilates are made taller and wide dilates are made shorter.)

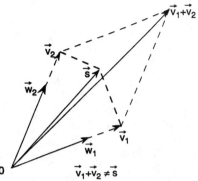

Figure 5. In a non-orthogonal basis, computing the scalar products of the signal and the basis functions does not allow us to reconstruct the original signal. The two basis functions $\vec{w}_1$ and $\vec{w}_2$ are not orthogonal. Projecting the signal $\vec{s}$ onto the axes gives two new vectors, $\vec{v}_1$ and $\vec{v}_2$, whose sum does *not* equal $\vec{s}$.

Non-Orthogonal Bases

To appreciate the convenience of being able to compute a wavelet co-efficient with a single scalar product, let's see what happens when we decompose a signal in a basis that is discrete but not orthogonal. Computing the scalar products of the signal and the basis functions no longer gives us coefficients that allow us to reconstruct our original signal. As shown in Figure 5, the scalar product of the signal $\vec{s}$ and the basis function $\vec{w}_1$ still gives a coefficient c_1, and the scalar product of $\vec{s}$ and $\vec{w}_2$ gives a coefficient c_2. But $c_1\vec{w}_1 + c_2\vec{w}_2$ equals $\vec{v}_1 + \vec{v}_2$, which does *not* equal the signal $\vec{s}$.

If we want to be able to reconstruct the original signal, we can create vectors $\vec{v}_1$ and $\vec{v}_2$ such that $\vec{v}_1 + \vec{v}_2 = \vec{s}$, but the coefficients needed to express $\vec{v}_1$ and $\vec{v}_2$ in terms of $\vec{w}_1$ and $\vec{w}_2$ can no longer be computed with a single scalar product. When, as shown in Figure 6, we draw the (non-perpendicular) projection from $\vec{s}$ to $\vec{w}_1$, we have to make it parallel to $\vec{w}_2$. When we draw the (non-perpendicular) projection from $\vec{s}$ to $\vec{w}_2$, we have to make it parallel to $\vec{w}_1$: computing a single coefficient involves all the vectors of the basis. In two dimensions this isn't bad; computing two coefficients requires solving two simultaneous linear equations in two variables. But in n dimensions, one has to solve n simultaneous linear equations in n variables: roughly n^3 operations.

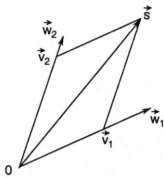

Figure 6. This non-orthogonal decomposition does allow us to reconstruct the signal: $\vec{v}_1 + \vec{v}_2 = \vec{s}$. But when we draw the projection from $\vec{s}$ to $\vec{w}_1$, we have to make it parallel to $\vec{w}_2$: computing a single coefficient involves all the vectors of the basis. In higher dimensions this becomes difficult: in n dimensions, we would have to solve n simultaneous linear equations in n variables—roughly n^3 operations.

The fact that all the vectors in a non-orthogonal basis come into play for the computation of a single coefficient is also bothersome when one wants to compute or adjust quantization errors (roughly, round-off errors).[4] In an orthogonal basis one can (by the theorem of Parseval) calculate the "energy" of the total error by adding the energies of the errors for each coefficient; it's not necessary to reconstruct the signal. And the error for each coefficient is independent of all the others. In a non-orthogonal basis, one has to reconstruct the signal to measure the error, and trying to adjust the "sub-errors" is tricky; as soon as one touches one coefficient, all the other coefficients are affected.

---

[4]This use of the word *quantization* is somewhat unfortunate, as the word has a very different meaning in quantum mechanics; *quantizing* a harmonic oscillator, for example, means turning it into a quantum mechanical system. In signal processing, quantization means approximating numbers with very many decimals by assigning them the closest number from a predetermined set. If that set is the integers or the numbers that stop after two digits, for example, then quantization is the same as rounding off. "But in most applications the quantization levels, which is what the numbers are called, are not equally spaced, and quite some work and real understanding of the problem go into a good choice of the quantization levels in order to get the best results," says Ingrid Daubechies of Princeton University.

More on Redundancy

Talking about redundancy can be a little confusing. The most dramatic comparison is between orthonormal transforms and continuous transforms where, as Yves Meyer puts it, "everything is said ten times." In an orthonormal basis, each vector encodes information that is encoded nowhere else; if a single vector of the transform is removed, information is lost. This conciseness can be desirable, but if information is very important or costly, entrusting it to an orthonormal transform may be risky. In a continuous transform, "if there's a misprint it doesn't matter," says Meyer. "But if there are mistakes in a short text, like a Chinese poem, you're lost."

But an orthonormal transform is just a special case of discrete transforms, which vary in how concise they are. Some discrete, non-orthogonal transforms are equally concise. In addition, an orthonormal transform can still contain redundancies that come from statistical correlations of the signal, as mentioned in the discussion of Best Basis, p. 96.

Scalar Products of Complex-Valued Vectors

We should mention that when complex-valued vectors are used, as they are in the proof of the Heisenberg uncertainty principle on p. 281, the definition of scalar product is slightly different. If the coordinates of $\vec{a}$ and $\vec{b}$ are complex numbers, then

$$\left\langle \vec{a}, \vec{b} \right\rangle = \left\langle \begin{bmatrix} a_1 \\ a_2 \\ \vdots \\ a_n \end{bmatrix}, \begin{bmatrix} b_1 \\ b_2 \\ \vdots \\ b_n \end{bmatrix} \right\rangle = a_1\bar{b_1} + a_2\bar{b_2} + \ldots + a_n\bar{b_n},$$

where $\bar{b_n}$ is the *complex conjugate* of b_n—the number obtained by changing the sign of the imaginary part of b_n.

Multiresolution

With multiresolution theory, Stéphane Mallat linked orthogonal wavelets with the filters used in signal processing. In this approach, the wavelet is upstaged by a new function, the *scaling function*, which gives a series of pictures of the signal, each at a resolution differing by a factor of two from the previous resolution. In one direction, these successive images approximate the signal with greater and greater precision, approaching the original. In the other direction, they approach zero, containing less and less information.

The wavelets still have an important role to play. They encode the difference of information between two resolutions: the details that must be added to one picture of the signal to obtain the picture at the resolution twice as great.

The idea of analyzing a picture at different resolutions was commonplace in image processing when Mallat made the connection with wavelets. Since an image generally contains structures of very different sizes, there is no single optimal resolution for analyzing them. "A multiresolution decomposition," he wrote, "enables us to have a scale-invariant interpretation of the image" [Mallat 3, p. 674]: that is, an interpretation that does not depend on the distance between the image and the "camera." With multiresolution, it's as if one brought the camera closer to capture details of the image and then moved it back to see the overall structure.

The various multiresolution decompositions that existed in image processing (Haar, cubic splines, cardinal sine ...) used a scaling function to go from one resolution to the next. Independently, working with wavelets,

Yves Meyer began searching for such a function. The wavelets he had made formed an orthogonal basis, as long as one dilated the "mother" wavelet without limit in both directions, both stretching and compressing it.

Meyer thought a function should exist that would give a starting point for an orthogonal decomposition; such a function and its translates would take care of all the low-frequency information, leaving the high frequencies to the wavelets. He and his student Pierre Gilles Lemarié (now Lemarié-Rieusset) found this function [Lemarié, Meyer] and another for Lemarié's own basis of spline functions. Lemarié called these scaling functions the *miracle of the low frequencies* [Lemarié, p. 31].

"I initially worked on the parallel between wavelets and the multiresolution idea used in image processing, and that's why I contacted Yves Meyer," remarks Mallat. "However, his own path had led him not far from this point; that's why everything went very fast when we met. It is hard to say precisely who did what between us; it was the result of a joint work, with different perspectives."

Filters

Mathematicians classify functions in all kinds of ways. The viewpoint of signal processors is simpler: to them, a function is either a signal to be analyzed or a filter with which to analyze a function. A classical filter is an electrical circuit with one wire that carries the signal in and another wire that carries the filtered signal out. But a filter can also be a function (or, if it is digital, a sequences of numbers). The effect of a filter, whether physical or abstract, is easier to understand in Fourier space: the Fourier transform of the signal is multiplied by the Fourier transform of the filter, letting certain frequencies pass through while blocking others.

If, for example, the Fourier transform of the filter is almost 1 near zero, and almost 0 everywhere else, as shown in Figure 1, the signal's low frequencies will survive this multiplication by 1, but the high frequencies will be essentially eliminated. This is a low-pass filter. The result in "physical" space is to smooth the signal: the small variations given by the high frequencies disappear, leaving the general tendency. (Pushing

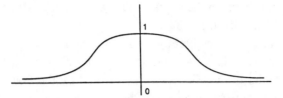

Figure 1. The effect of a filter is best understood in Fourier space: the Fourier transform of the signal is multiplied by the Fourier transform of the filter. Above is the Fourier transform of a low-pass filter, almost 1 near 0 and almost 0 everywhere else. The signal's low frequencies will survive this multiplication by 1, but the high frequencies will be essentially eliminated. The result in "physical" space is to smooth the signal.

the "high fil" button on an amplifier for a stereo activates a low-pass filter, eliminating high-frequency scratches as well as some of a soprano's vibrato.)

Mallat realized that one can incorporate wavelets into a system that uses a cascade of filters to decompose a signal. Each resolution has its own pair of filters: a low-pass filter associated with the scaling function, giving an overall picture of the signal, and a high-pass filter associated with the wavelet, letting through only the high frequencies associated with the variations, or details. The two filters complement each other; that which one blocks, the other lets through. (Of course "high" or "low" frequencies are relative. The low frequencies encoded by the low-pass filter at a fine resolution may be higher than the high frequencies encoded by the high-pass filter at a coarse resolution.)

Incorporating orthogonal wavelets in a multiresolution system based on filters made it possible to have an orthogonal multiresolution: fast, complete, concise. The pyramid algorithms of Burt and Adelson doubled the size of an image; Mallat's multiresolution computes the same number of coefficients as the number of pixels in the original picture. Image processing also gained because wavelets introduced the concept of regularity and gave a solid mathematical foundation to existing algorithms.

Wavelets gained as well: in this scheme, computing wavelet coefficients requires nothing more than a series of simple calculations involving short digital filters. The wavelet transform—some argue that *fast wavelet transform* is a tautology—immediately became a serious rival to the FFT for a whole range of applications. (The algorithm is described in *The Fast Wavelet Transform*, p. 183.)

In addition, the same filters gave a systematic way to construct new orthogonal wavelets. The scaling function is not the father of the wavelets, nor is the wavelet the mother; both are created by the Fourier transform (also called the *transfer function*) of the low-pass filter.

By judiciously choosing this function, one can make customized wavelets with the desired properties; a trigonometric polynomial, for example, generates a scaling function with compact support and (at least for one-dimensional wavelets) wavelets with compact support. Orthogonal wavelets—rare, sought-after, and more than a little mysterious—became members of an infinite family of functions, well enough understood so that one can talk nonchalantly about making one's own orthogonal wavelets to order.

And although Mallat didn't think in these terms, his theory also makes it possible to think about wavelets geometrically. In this perspective, the transfer function defines a curve on the surface of a sphere in four-dimensional space (more precisely, a parametrized curve on the surface of the sphere with radius 1 centered at the origin, in complex two-dimensional space—a *3-sphere*).

Each multiresolution corresponds to such a curve, and all these curves join the points (1,0) and (0,1) on the sphere. One can then "draw" (we are, don't forget, in four dimensions) almost any curve that joins those two points to create a scaling function and wavelet. If this approach may seem superfluous when talking about one-dimensional wavelets, some researchers think it could be useful when working in two or more dimensions. For two-dimensional wavelets, rather than a curve on a 3-sphere we have a parametrized torus on a 7-sphere; for three-dimensional wavelets, it's a 15-sphere. The construction of wavelets then comes up against topological obstacles.

Mallat foresaw none of these ramifications for wavelets when he first read Meyer's paper describing his orthogonal wavelets:

> Intuitively it was quite clear that there was a relation between Meyer's wavelets and the image processing pyramids. My ambition was to understand that relation in order to apply the mathematical results to image

processing. I had much too much respect for mathematics to think that I could contribute to the mathematical sides, and I still had this clear vision of the flow of knowledge going from pure mathematics to applications and not the reverse. It is when trying to establish the relation that I realized that the image processing approach could contribute to the mathematical understanding of wavelets.

The Definition of a Multiresolution

A multiresolution analysis can be compared to approximating the number 87/7 using decimals: 12.4285714 .... The scaling function, compressed or dilated, gives an image of the signal at a given resolution, just as one can round off 87/7 as 10; 12; 12.4; or 12.42 ...depending on the desired accuracy. Wavelets encode the difference of information between two resolutions (for wavelets, resolutions differing by a factor of two; for decimals, by a factor of 10). Between 10 and 12 the wavelets encode the detail 2; between 12 and 12.4, other, smaller wavelets encode the detail 0.4; between 12.4 and 12.42, still others encode 0.02, and so on.

The further one descends in levels of details, the more accurate the approximation. In the other direction, stretching the scaling function more and more, we end by seeing nothing at all, as if we were trying to approximate 87/7 using hundreds. At that point all the information has been treated like details encoded by the wavelets: $10 + 2 + 0.4 + 0.02 + 0.008 + \cdots$.

Our decimal system can approximate any number without redundancy and with arbitrary precision; a multiresolution can do the same for any signal, if it satisfies four conditions. (For a more detailed but still relatively elementary discussion, see [Strichartz].) These conditions are:

(1) *The scaling function must be orthogonal to its translates by integers.*

When we translate the scaling function φ ("phi") by a whole number, the translated functions must all be orthogonal to each other: the scalar product of φ and any of its translates is zero.

The condition of orthogonality to translates by integers is easy to see in the case of the scaling function corresponding to the Haar function, shown in Figure 2. The Haar function, which dates from 1910 [Haar],

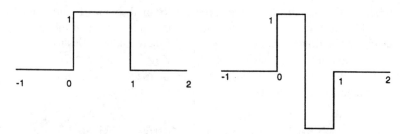

Figure 2. Left: the Haar scaling function. Right: the Haar function, the first and simplest wavelet generating an orthogonal family of wavelets. The Haar scaling function and wavelet provide a useful example for understanding the multiresolution theory of Mallat and Meyer.

is the first wavelet generating an orthogonal family of wavelets. It is generally considered too jerky to be useful in image processing; when an image is encoded with Haar, the discontinuities create artifacts.

The scaling function corresponding to the Haar wavelet has the value 1 for $0 \leq x < 1$, and 0 for all other values of x. If we translate it by an integer, wherever the scaling function has the value 1, its translate has the value 0, and their scalar product is zero: they are orthogonal to each other. It is harder to satisfy this condition for functions with a support greater than 1. When we translate such a function by 1, the two functions overlap, and we need delicate cancellations of positive and negative terms in order to avoid correlation.

(2) *The signal at a given resolution contains all the information of the signal at coarser resolutions.*

This is expressed by talking of the spaces V_j (i.e., the spaces $\ldots, V_{-2}$, $V_{-1}, V_0, V_1, V_2, \ldots$). The space V_0 is by definition the space generated by the scaling function and all its translates by integers; everything that can be expressed by these functions is contained in V_0, and everything that is contained in V_0 can be expressed by these functions. An example of a function in the space V_0 of Haar is given in Figure 3.

The space V_j is formed with the functions of V_0 compressed by a factor of 2^j. For example, for the Haar multiresolution, V_1 is the space of bar graphs with possible discontinuities at the half-integers: it is created by the scaling function compressed by a factor of two and translated by half-integers.

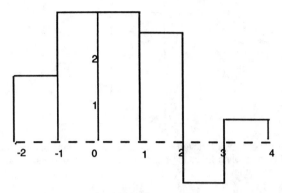

Figure 3. The space V_0 of the Haar multiresolution is the space generated by the Haar scaling function and all its translates by integers. Above is one function in that space. All such functions are bar graphs—locally constant, with possible discontinuities at the integers. By compressing the Haar scaling function by a factor of 2^j one can form the space V_j. For example, the space V_1 of Haar contains locally constant functions with possible discontinuities at the half-integers. The space V_0 is contained in V_1, V_1 is contained in V_2, and so on.

The space V_0 of a multiresolution must be contained in V_1, which is obvious in the case of Haar: a bar graph with possible discontinuities at the integers can be written as a bar graph with possible discontinuities at the half-integers. And if V_0 is contained in V_1, then V_1 is contained in V_2, V_2 in V_3, etc. (Some authors, including Daubechies, use the opposite notation, in which V_1 is contained in V_0: one more trap for the unwary.)

Once again the Haar example can be misleading. One can translate an arbitrary function by integers and compress it by a factor of two; there's no reason to think that the spaces V_j created by the function and its translates and dilates will necessarily be nested in each other. Consider a function analogous to the Haar scaling function, with the value 1 between $\frac{1}{4}$ and $\frac{3}{4}$, and 0 elsewhere, shown in Figure 4.

If we translate it by integers we get a space V_0 of functions that are bar graphs with the bars separated; an example is given in Figure 5. If we compress it by a factor of two, the function will have the value 1 between $\frac{1}{8}$ and $\frac{3}{8}$, and 0 elsewhere. Translating this new function by half-integers gives us a new space V_1, also of bar graphs, an example of which is shown in Figure 6. But in this case the space V_0 is not contained in V_1; our new function could not be used to create a multiresolution system.

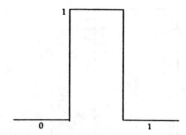

Figure 4. The above function is analogous to the Haar scaling function, but unlike the Haar scaling function, it cannot be used to create spaces V_j that are nested in each other.

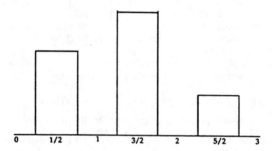

Figure 5. An example of a function in the space V_0 created by the integer translates of the function in Figure 4.

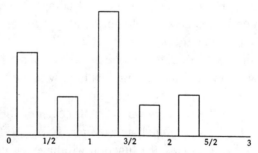

Figure 6. A function in the space V_1 created by the function in Figure 4 compressed by a factor of 2 and by its half-integer translates. The space V_1 does not contain the space V_0 shown in Figure 5. The function shown in Figure 4 cannot be used to create a multiresolution system: a "high-resolution" image of the signal would not contain all the information of a "low-resolution" image.

(3) *The function 0 is the only object common to all the spaces V_j.*

If we dilate the scaling function sufficiently (using the word "dilate" here in its proper sense, to make wider), eventually the image of the signal given by the scaling function will be completely drained of information, as if we were trying to approximate $87/7 \approx 12.428$ using hundreds:

$$\lim_{j \to -\infty} V_j = \bigcap V_j = \{0\}.$$

(4) *Any signal can be approximated with arbitrary precision:*

$$\lim_{j \to \infty} V_j = L^2(\mathbb{R}).$$

Constructing a Multiresolution

If these four conditions are satisfied, then there exists a wavelet that, with its translates by integers and dilates by a factor of two, can encode the difference of information between the signal seen at two successive resolutions. In other words, the space W_j associated with the wavelet is orthogonal to the space V_j and represents the difference between V_j and V_{j+1}:

$$W_j \oplus V_j = V_{j+1}.$$

It is not *a priori* obvious that functions satisfying these conditions exist, especially conditions 1 and 2 (other than the Haar scaling function). But Mallat found that a scaling function and its wavelet can be constructed using the Fourier transform—the *transfer function*—of an almost arbitrary filter. This immediately gives rise to an infinite family of multiresolution systems, each with its own scaling function and "mother" wavelet.

We will sketch the procedure here; for more detail, see [Strichartz] (also [Daubechies], [Strang], and [Mallat 3]). We will start backwards, with a scaling function, and find the associated transfer function. To do this, we write the scaling function as a combination of its translates at a resolution twice as fine, each translate multiplied by a coefficient (this uses condition 2):[1]

$$\varphi(x) = \sum_{n=-\infty}^{\infty} a_n 2\varphi(2x - n).$$

(We write 2φ rather than φ to keep the integral of these new functions equal to 1. Since they are compressed by a factor of 2, they have to be twice as tall.)[2] For Haar we get:

$$\varphi(x) = \frac{1}{2}\Big(2\varphi(2x)\Big) + \frac{1}{2}\Big(2\varphi(2x - 1)\Big),$$

as shown in Figure 7.

So for Haar, $a_0 = a_1 = \frac{1}{2}$, and the other a_n are zero.

Next we create a Fourier series A using these same numbers a_n as coefficients:

$$A(\xi) = \sum_{n=-\infty}^{\infty} a_n e^{2\pi i n \xi} \tag{7.1}$$

where A satisfies the conditions $A(0) = 1$ and

$$|A(\xi)|^2 + \left| A\left(\xi + \frac{1}{2}\right)\right|^2 = 1. \tag{7.2}$$

We also impose a somewhat weak condition of regularity: the a_n's tend to zero reasonably fast. (The function A is periodic of period 1.

---

[1] As discussed in [Strichartz, p. 543], this formula says that φ is a fixed point of the linear transformation S given by

$$(S(f))(x) = \sum_{n=-\infty}^{\infty} 2a_n f(2x - n),$$

and one way to find a function φ for a given a_n is to iterate S. This leads to a construction very similar to iterated function systems. We will discuss instead Mallat's solution via Fourier transforms and infinite products, inspired by cascades of filters.

[2] It would be more usual to keep the *lengths* of the functions—and therefore the integral of the functions *squared*—equal to 1, as in [Daubechies]. For this one would write $a_n\sqrt{2}\,\varphi(2x-n)$ in the above formula, and the coefficients would then be $a_0 = a_1 = \frac{1}{\sqrt{2}}$. Using $\frac{1}{2}$ will make our computations simpler.

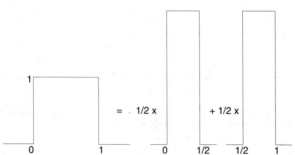

Figure 7. The Haar scaling function expressed in terms of its half-integer translates at the next finer resolution.

Mallat uses period 2π, but other than that his equation 17b [Mallat 3, p. 678] is the same.)

From a geometric point of view, the condition in formula (7.2) says that the curve whose first coordinate is the complex number $A(\xi)$ and whose second coordinate is $A(\xi + \frac{1}{2})$ lives on the 3-sphere of radius 1. In terms of image processing, condition (7.2) corresponds to a condition for the low-pass filter of a pair of complementary twin filters, saddled with the barbaric name *quadrature mirror filters*. (The function A was first studied in the context of filters by Mintzer and Smith and Barnwell, [Mintzer], [Smith, Barnwell], before Mallat rediscovered it in the context of wavelets.)

The function A is in fact the Fourier transform of the low-pass digital filter (we'll call it a) that consists of the sequence of numbers $\ldots, a_{-1}, a_0, a_1, \ldots$. (If the statement that A is the Fourier transform of a bothers you, see *The Fourier Transform of a Periodic Function*, in the Appendix, p. 285.)

Consider the values $A(0)$ and $A(\frac{1}{2})$. Since by definition $A(0) = 1$, equation (7.2) gives us $A(\frac{1}{2}) = 0$. This gives us a graph (Figure 8) that we can recognize—at least the central portion—as the Fourier transform of a low-pass filter. (Although we don't justify this precision here, if A is nonzero between $-\frac{1}{4}$ and $\frac{1}{4}$, there exists a corresponding scaling function. The converse is almost true: if A is created from a scaling function, A is nonzero in a suitably modified version of the interval $[-\frac{1}{4}, \frac{1}{4}]$, which we won't describe here.)

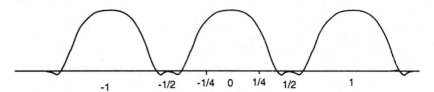

Figure 8. The function A, used to create scaling functions and wavelets, is the Fourier transform of the low-pass digital filter a.

The Haar Multiresolution

Let's see what this gives for Haar. Creating a Fourier series A_H (H for "Haar") with our coefficients $a_0 = a_1 = \frac{1}{2}$, we get:

$$A_H(\xi) = \frac{1}{2}\left(e^{0(2\pi i \xi)} + e^{1(2\pi i \xi)}\right) = \frac{1}{2}\left(1 + e^{2\pi i \xi}\right).$$

Using the formula $\cos\theta = \frac{1}{2}(e^{i\theta} + e^{-i\theta})$, we get:

$$A_H(\xi) = e^{\pi i \xi}\cos\pi\xi$$

and

$$A_H\left(\xi + \frac{1}{2}\right) = e^{\pi i(\xi + \frac{1}{2})}\cos\pi\left(\xi + \frac{1}{2}\right) = e^{\pi i \xi}(-i\sin\pi\xi).$$

(The last step uses the formulas $\cos\left(\theta + \frac{\pi}{2}\right) = -\sin\theta$ and $e^{\frac{\pi i}{2}} = i$.)
We see that the condition of formula (7.2) is satisfied:

$$\left(\underbrace{\left|e^{\pi i \xi}\right|}_{=1}\,|\cos\pi\xi|\right)^2 + \left(\underbrace{\left|e^{\pi i \xi}\right|}_{=1}\,|-i\sin\pi\xi|\right)^2 = 1.$$

Creating a Scaling Function

In practice we want to go in the other direction. It's not hard to create a function A that satisfies our conditions. When Mallat saw the connection between wavelets and filters, an extensive mathematical literature already existed on these filters and on numerical ways of constructing them. But even if the Fourier coefficients of A are also the coefficients of a scaling function expressed as the sum of its translates at a resolution twice as fine, we don't know what this scaling function looks like.

Mallat found a way of going from the function A to the Fourier transform of the scaling function, $\hat{\varphi}$:

$$\hat{\varphi}(\xi) = \prod_{j=1}^{\infty} A\left(\frac{\xi}{2^j}\right). \tag{7.3}$$

This formula (formula 18, [Mallat 3, p. 678] and formula 5.10, [Strichartz, p. 548]) is an infinite product. To compute the value of $\hat{\varphi}$ at each ξ, we multiply together all the values of A evaluated at the points $\xi/2^j$, for j from 1 to infinity. (These multiplications correspond in "physical" space to a cascade of convolutions of the low-pass filter with itself at different scales.) The idea of an infinite product may be intimidating, but one shouldn't take infinity too seriously here; the product converges at dizzying speed towards the limit, and Fourier transforms of scaling functions are routinely constructed with as few as six terms in the product.

To find the scaling function, we invert (undo) the Fourier transform. When the number of nonzero coefficients is finite, we get a scaling function with compact support. (This is the case for Haar and for the scaling functions corresponding to the Daubechies wavelets.)

Making Wavelets

Where are wavelets in all this? In the geometrical perspective, we start by creating a second curve on the 3-sphere. Each point of the first curve (the curve associated with A) is a vector; we build the wavelet curve with vectors that are orthogonal (perpendicular) to the vectors of the curve associated with A. We use this new curve to construct a function that we will call D (for "differences," since wavelets encode differences), just as

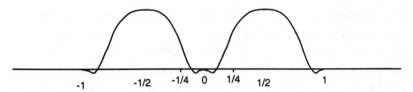

Figure 9. The graph of the function D looks like the Fourier transform of a high-pass filter. Orthogonal wavelets can be created from the functions A and D, using formula (7.4).

we constructed A from the first curve: the first coordinate of this new curve is $D(\xi)$ and the second $D(\xi + \frac{1}{2})$.

There exists more than one curve orthogonal to the first curve, so the same function A can generate more than one wavelet, but there is a nicest one: there is a rule associating to a curve another curve orthogonal to it so that if the scaling function corresponding to the first curve is real (as opposed to complex), the wavelet associated to the second curve will also be real. In addition, if the scaling function has compact support, this "nice" curve produces a wavelet with compact support, which isn't necessarily true of the other curves. (In two or more dimensions, this rule doesn't work.)

The wavelet ψ is created from D and A using the formula:

$$\hat{\psi}(\xi) = D\left(\frac{\xi}{2}\right) \prod_{j=2}^{\infty} A\left(\frac{\xi}{2^j}\right). \tag{7.4}$$

If the geometrical approach isn't appealing, one can think of D as the Fourier transform of a high-pass filter. The condition $|D(\xi)|^2 + |D(\xi + \frac{1}{2})|^2 = 1$, which can be interpreted as saying that the curve associated with D lives on the 3-sphere, is the same condition as condition (7.2) for the function $A(\xi)$: D has the same shape as A. But since the two curves are orthogonal, wherever $A(\xi) = 1$, $D(\xi) = 0$, and vice versa; the graph of $D(\xi)$ looks like the Fourier transform of a high-pass filter, as shown in Figure 9.

One could object that since A and D are periodic, the Fourier transforms of the "low-pass" and "high-pass" filters alternate (Figure 10), which seems to make nonsense of the notion of "high" and "low" frequencies.

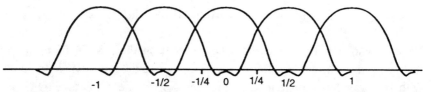

Figure 10. The functions A and D—the Fourier transforms of the low-pass and high-pass filters—alternate. But since the frequencies of a digital filter repeat, wrapping around a circle, one considers only, for example, the region between $-\frac{1}{2}$ and $\frac{1}{2}$.

But the frequencies of a digital filter (like those of a sampled signal) don't live on a line from minus infinity to infinity; they wrap around a circle. That is, the Fourier transform of a sampled signal is periodic. To compute the Fourier coefficients of a signal sampled with period p, we multiply the sampled values of the signal by the complex exponential of the frequency in question, add them, and multiply by the sampling period p (as we did in equation (4.2) in *The Fast Fourier Transform*, page 143). The Fourier coefficient for frequency $\tau + \frac{1}{p}$ will be the same as that for frequency τ:

$$p \sum_{n=-\infty}^{\infty} f(np)\, e^{2\pi i(\tau+\frac{1}{p})np} = p \sum_{n=-\infty}^{\infty} f(np)\, e^{2\pi i\tau np} \underbrace{e^{2\pi in}}_{=1}.$$

So the coefficients of $\hat{f}$ at τ, $(\tau + \frac{1}{p})$, $(\tau + \frac{2}{p})$... are all equal. (We approach this same issue from a different point of view in *The Fourier Transform of a Periodic Function*, p. 285, in which we use distributions to justify the statement that the periodic function A is the Fourier transform of the sequence a_n; the a_n can of course be thought of as a sampled signal.)

When we talk about band-limited signals sampled at the rate imposed by Shannon, the terms "high" and "low" frequencies keep their ordinary meanings. For example, if the bandwidth is 10,000 hertz and we sample 20,000 times a second, we choose our units of time so that 1 in the τ-coordinate (horizontal coordinate) of the above figure corresponds to 20,000 hertz, and we consider only the region between $-\frac{1}{2}$ and $\frac{1}{2}$.

The Scaling Function as Father

We can also obtain the wavelet ψ directly from the scaling function φ, justifying the expression "father function," since the space W associated with the wavelet is contained in the space V at the resolution twice as fine: $V_0 \oplus W_0 = V_1$. Then we have

$$\psi(t) = 2 \sum_{n=-\infty}^{\infty} d_n \varphi(2t - n). \qquad (7.5)$$

In the case of Haar we know that

$$\psi(t) = \varphi(2t) - \varphi(2t - 1) = \begin{cases} 1, & 0 \le t < 1/2, \\ -1, & 1/2 \le t < 1, \\ 0, & \text{otherwise.} \end{cases}$$

So $d_0 = \frac{1}{2}$, $d_1 = -\frac{1}{2}$ and the other d_n are zero. The ease with which a wavelet can be constructed from the scaling function stems from the intimate and simple relationship between the coefficients a_n of A and d_n of D. "It is absolutely remarkable," Strang writes [Strang, p. 294], that the d_n are identical to the a_n "but in reverse order and with alternating signs." (In the case of Haar, we get $d_0 = a_0$ and $d_1 = -a_1$; Strang is using slightly different labeling of the a_n.) Yet the simplicity of the relationship is not entirely surprising, considering how similar their graphs are, or given the fact that one can think of A and D as being constructed from two curves (on a 3-sphere) that are orthogonal to each other.

Computing a Wavelet Transform Without Wavelets

A curious consequence of multiresolution is that can we transform a signal into wavelets using neither wavelets nor scaling functions. To compute the wavelet transform all we need are filters. Rather than taking the scalar product of the scaling function (or the wavelet) with the signal, we *convolve* the signal with these filters (see *The Fast Wavelet Transform*, p. 183).

"It is amazing to compute with a function we do not know," writes Strang [Strang, p. 295]. But "when complicated functions come from a simple rule, we know from increasing experience what to do: Stay with the simple rule."

Note: The function A and the coefficients a_n here and in [Strichartz] are H and h_n in [Mallat 3]; the same coefficients are c_n in [Strang]; in [Lemarié], the function A (resp. H) is written m_0... In addition, not everyone normalizes the same way; our Haar low-pass filter consists of $a_0 = a_1 = \frac{1}{2}$, [Daubechies] uses $\frac{1}{\sqrt{2}}$, and [Strang] uses 1. *Caveat emptor*.

The Fast Wavelet Transform

Multiresolution theory gives a simple and fast method for decomposing a signal into its components at different scales. We progressively drain the signal of its information, beginning with small details and continuing on to larger and larger components, as shown in Figure 1. At each step we encode the "details" as wavelet coefficients and work at the next step with the signal seen at half the previous resolution.

In the language of wavelets, the scaling function is dilated to make an image of the signal at half resolution; in the language of signal processing, a low-pass filter is applied to the signal, and the result is subsampled. (Flouting etymology, signal processing texts speak of "decimating" the signal by a factor of two: taking one sample out of two.)

Since at each step we take only half as many samples as before, it doesn't take long before the signal is reduced to nothing, or almost nothing. The trick is that we don't lose anything: the information encoded by the wavelets is precisely the information subtracted from the signal when we "decimate" it. We can retrace our steps and find the original signal, adding to the picture of the signal at one resolution the information encoded by the wavelets at that resolution, in order to get the image of the signal at a resolution twice as fine.

We can think of the procedure as a series of additions and subtractions. In Figure 2, we begin with 32 samples of the signal.

We group them by pairs, computing the average and the difference for each pair. The first step gives 16 differences (the wavelet coefficients) and 16 averages (the coefficients of the scaling function). Next we work with the 16 averages, two by two, computing 8 wavelet coefficients and 8 averages, and so on.

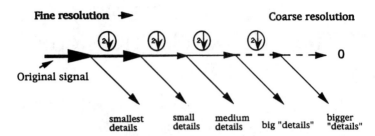

Figure 1. The fast wavelet transform works from fine resolution to coarse. At each resolution, the signal is analyzed with both wavelets and scaling function. The wavelets encode the details, while the scaling function produces an image of the signal at half resolution, taking one sample out of two. The process is repeated until nothing, or virtually nothing, is left.

"What makes it really fast is that I only handle small numbers of points at every computation," Daubechies says. At the fourth level of our pyramid, for example, each of the four average values represents the average of eight original points. This gives a linear transformation: our signal has $n = 32$ points; the total of all the additions and subtractions is $62 \approx 2n$. (The classical Fourier transform requires n^2 computations, and the FFT requires $n \log n$.) As we will see, the speed of the fast wavelet transform (FWT) depends on the kind of wavelet used, but the number of

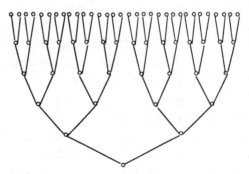

Figure 2. The fast wavelet transform can be thought of as a series of additions and subtractions. Here we start at the top with 32 samples of a signal, grouping them by pairs. The wavelet computes differences, which are saved as wavelet coefficients. The scaling function computes averages of the pairs, providing 16 samples that give an image of the signal at half-resolution. These are used as the next input.

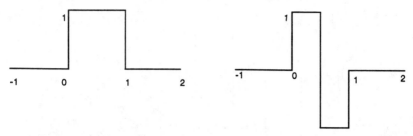

Figure 3. The Haar scaling function, left, and the Haar wavelet, right, provide a simple example of the fast wavelet transform.

computations required always increases linearly with the size of the signal to be transformed. Computing a fast wavelet transform of a signal with n points takes about $2cn$ computations, where c depends on the wavelet used (which must have compact support).

The transformation is concise. For a signal that consists of 32 samples, we end up with 32 coefficients: 31 wavelet coefficients and the last scaling function coefficient, which represents the average value of the signal. But by itself the wavelet transform does not compress the signal; if we want compression, we have to take advantage of the fact that for most signals, a given value is likely to be identical or very close to neighboring values, giving a difference—a wavelet coefficient—that is zero or very small.

The Haar Wavelet Transform

This example of additions and subtractions is what happens in the Haar wavelet transform, using the Haar scaling function and wavelet shown in Figure 3.

In Figure 4, the dotted lines indicate an approximation of a signal by the Haar scaling function and its translates. For each of four intervals (0 to 1, 1 to 2, 2 to 3, 3 to 4), these scaling functions give average values.

The scaling function coefficients are obtained by computing the integral of the product of the scaling function and the signal. Since the scaling function has the value 1 between 0 and 1, the product equals the signal on that interval. The integral for the interval [0,1] is .5. When we translate the scaling function by 1, the integral of the product of scaling

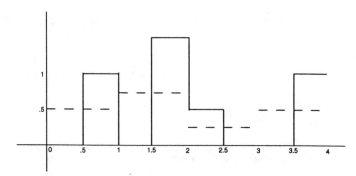

Figure 4. A signal and its approximation using the Haar scaling function (dotted lines). The value of the signal at 0.5 is 0, and at 1 it is 1, so the average for that interval is 0.5.

function and signal in the interval [1,2] is .75, and so on. Each scaling function coefficient gives the average value of the signal over a given interval.

The wavelets encode the difference (more precisely, half the difference) between two adjacent values of the signal. The wavelet has the value of 1 between 0 and .5 and −1 between .5 and 1. The product of the wavelet and signal for the first half of the interval [0,1] is 0, while the product for the second half is −1. This gives an integral (a wavelet coefficient) of −.5, or half the difference of the signal over [0,.5] and the signal over [.5,1].

Convolutions

We can think of the wavelet transform geometrically, the scaling function smoothing the signal and the wavelet picking out differences, but in practice the transform is computed with simple arithmetic. A digital signal is *convolved* with two digital filters. (Readers who know what a convolution is may wish to skip ahead to the next section, *Convolutions and the Wavelet Transform*, page 188.)

One can think of convolutions geometrically, but we will explain them in terms of arithmetic, since that is what computers do when they compute a wavelet transform. From this point of view a convolution is nothing more than the algorithm, learned in elementary school, which we use to multiply two numbers. But the interpretation is different. Instead of saying, for example, that we have multiplied 426 by 32 to get 13 632, we say that we have *convolved* the sequences of numbers [4,2,6] and [3,2] to get the sequence [1,3,6,3,2].

The convolution of the sequences a and b is written $(a \star b)$. The kth term of the sequence $a \star b$ is given by

$$(a \star b)_k = \sum_{j=-\infty}^{\infty} a_j b_{k-j}. \qquad (8.1)$$

If the a_j and b_j are nonzero only for $j \geq 0$ then

$$(a \star b)_k = \sum_{j=0}^{k} a_j b_{k-j},$$

and for the second entry of $(a \star b)$ we have

$$(a \star b)_2 = \sum_{j=0}^{2} a_j b_{2-j} = (a_0 b_2 + a_1 b_1 + a_2 b_0).$$

The connection with multiplication is clear if we write the sequences "backwards" the way we write numbers (inherited from the Arabs, who also read from right to left):

$$a_N, \ldots, a_2, a_1, a_0 = \ldots + 100a_2 + 10a_1 + a_0$$
$$b_M, \ldots, b_2, b_1, b_0 = \ldots + 100b_2 + 10b_1 + b_0$$

Finding $(a \star b)_2$ is the same as finding how many hundreds we have in the product of the two numbers $(a_N \ldots a_2 a_1 a_0)(b_M \ldots b_2 b_1 b_0)$. The term $a_0 b_2$ corresponds to multiplying the number of hundreds in b by the number of ones in a; the number $a_1 b_1$ corresponds to multiplying the number of tens in b by the number of tens in a, and so on. The only difference is that in a convolution we do "carries" only at the end, when we add up all the terms.

Convolutions and the Wavelet Transform

To transform a signal into wavelets, at each resolution we convolve the sampled signal with the sequence of numbers $\ldots, a_0, a_1, a_2, \ldots$ (the low-pass filter associated with the scaling function) and the sequence $\ldots, d_0, d_1, d_2, \ldots$ (the high-pass filter associated with the wavelet). These filters are discussed in *Multiresolution*, page 165. The a_n's and d_n's are called the *filter coefficients*, but this can be misleading; they are the filters.

We can now look at our Haar example in terms of convolutions.[1] The low-pass filter corresponding to the Haar scaling function consists of the two numbers $a_{-1} = a_0 = .5$. (In *Multiresolution* we called them a_0 and a_1, but our indices will give us trouble if we use that notation here.) We will use formula (8.1), thinking of a as the filter and b as our signal with eight values shown in Figure 4. (The value of the signal over $[0,.5]$ is $b_0 = 0$, its value over $[.5, 1]$ is $b_1 = 1$, etc.) We get four scaling function coefficients:

$$(a \star b)_0 = a_{-1}b_1 + a_0b_0 = .5(1 + 0) = .5;$$
$$(a \star b)_2 = a_{-1}b_3 + a_0b_2 = .5(1.5 + 0) = .75;$$
$$(a \star b)_4 = a_{-1}b_5 + a_0b_4 = .5(0 + .5) = .25;$$
$$(a \star b)_6 = a_{-1}b_7 + a_0b_6 = .5(0 + 1) = .5.$$

We can see that this corresponds to taking the average value of two adjacent samples. The eight original values of our signal give us four scaling function coefficients, corresponding to a smoothed signal.

The high-pass filter corresponding to the Haar wavelet has the values $d_{-1} = -.5$ and $d_0 = .5$. So the first wavelet coefficient is:

$$(d \star b)_0 = d_{-1}b_1 + d_0b_0 = -.5(1) + .5(0) = -.5,$$

and the second is

$$(d \star b)_2 = d_{-1}b_3 + d_0b_2 = -.5(1.5) + .5(0) = -.75.$$

---

[1]For an explanation of why it is true in general that convolving the signal with the low-pass and high-pass filters produces a wavelet transform, see for example [Daubechies, p. 156].

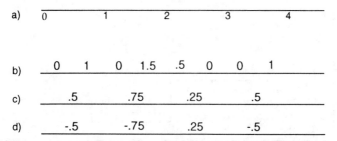

Figure 5. The original signal can be obtained by adding together the wavelet coefficients and scaling function coefficients at the first resolution. a) the x-axis; b) the sampled signal shown in Figure 4; c) the scaling function coefficients for the first-resolution approximation shown in Figure 4; d) the corresponding wavelet coefficients.

In Figure 5 we list the sampled values of the signal and the coefficients computed at this first resolution.

We can see that the wavelet coefficients give the information we would have to add to the scaling function coefficients in order to reproduce the original signal:

$$-.5 + .5 = 0,$$ the value of the signal at 3.5;
$$.25 + .25 = .5,$$ the value of the signal at 2.5;
$$-.75 + .75 = 0,$$ the value of the signal at 1.5, and so on.

At each resolution, the wavelet coefficients give the information we would have to add to the scaling function coefficients in order to equal the signal at the next higher resolution.

At the next resolution, shown in Figure 6, we start with the smoothed signal (the four scaling function coefficients) above and smooth it further (using a scaling function dilated to be twice as wide, and using wavelet dilates to encode the differences).

More Complicated Wavelets

The complexity of the computations increases linearly with the number of nonzero values of the filter—the number of *taps*. Convolving a signal with

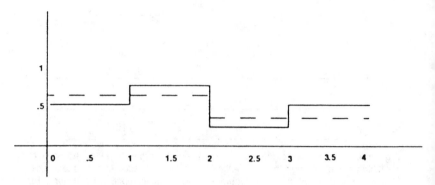

Figure 6. At the next resolution, the smoothed signal of Figures 4 and 5 is taken as the new starting point. Solid lines indicate the smoothed signal, dotted lines indicate its approximation by the dilated scaling function.

a filter that has only two nonzero values, a_0 and a_1, is like multiplying it by a two-digit number; $(a \star b)_k$ always has only two nonzero terms. For example, for $k = 3$ we get $a_0 b_3 + a_1 b_2$; for $k = 4$, we get $a_0 b_4 + a_1 b_3$, etc. This is the case for the Haar wavelet transform; it takes $2(2n)$ computations to compute the Haar wavelet transform of a signal with n values.

If a filter has three nonzero values, a_0, a_1, and a_2, then each $(a \star b)_k$ has three nonzero terms (for example, for $k = 4$ we have $a_0 b_4 + a_1 b_3 + a_2 b_2$). In this case a wavelet transform requires $2(3n)$ computations.

Wavelets with compact support all correspond to filters with a finite number of taps. But the greater the complexity of the wavelet (the more vanishing moments it has, for example), the more taps its filter will have, the greater the support of the wavelet, and the more one must work to compute each wavelet coefficient. On the other hand, Daubechies says, "the more vanishing moments, the better the local averages model smooth functions, at every generation, so the smaller the wavelet coefficients will be, which means that the quality of the approximation you make when you discard say half of all the coefficients—the smallest one—will be much better. In other words, for the same quality in the end you need to retain many fewer coefficients." Whether this makes up for the loss in speed depends on the application.

Which is Faster, the FFT or FWT?

At first glance, the fast wavelet transform (FWT) looks even faster than the FFT: roughly cn computations compared to $n \log n$. When n is very large, $n \log n$ is very much bigger than cn.

But "it's ridiculous, from the point of view of encoding, to compare a wavelet transform with an FFT of an entire image or signal," says Olivier Rioul of the Ecole Nationale Supérieure des Télécommunications in Paris. "To encode an image, for example, one cuts it into little blocks of 8 x 8 pixels and applies the FFT, or rather the DCT [discrete cosine transform], which has better properties, on each little block. So the comparison with the FWT is more complicated, especially since a wavelet transform is slightly more complex to implement."

Wavelets in Two Dimensions

When wavelets are used to encode two-dimensional signals (pictures, for example), often this is done by using "separable products" of a one-dimensional wavelet and a one-dimensional scaling function. This makes it possible to use the fast wavelet transform.

In this case, three two-dimensional wavelets, $\psi_1, \psi_2,$ and ψ_3, are constructed by multiplying together a one-dimensional scaling function ϕ and the corresponding wavelet ψ:

$$
\begin{aligned}
\psi_1(x,y) &= \phi(x)\,\psi(y), \\
\psi_2(x,y) &= \psi(x)\,\phi(y), \\
\psi_3(x,y) &= \psi(x)\,\psi(y).
\end{aligned}
$$

Each new wavelet measures variations in the image along a different direction: ψ_1 responds to variations in the vertical direction (horizontal edges, for example), ψ_2 responds to variations in the horizontal direction, and ψ_3 responds to variations along diagonals. The wavelets are dilated by scaling factors 2^j along x and y and translated on an infinite rectangular grid $(2^j n, 2^j m)$, where n and m are integers, in order to construct an orthonormal basis.

The fast wavelet transform of images is then calculated essentially by applying the one-dimensional wavelet transform along the rows and columns of the image. (For more on two-dimensional wavelet transforms, see for example [Mallat 4], [Strang, Nguyen], or [Vetterli, Kovacevic].)

193

Orientable Wavelets

The above method is fast, but favors the horizontal and vertical directions; "if you want a subband that is oriented at 30 degrees (instead of 0 or 90) you are out of luck," says Ted Adelson of MIT. If the goal is analysis rather than image compression, than it's desirable to forego this computational simplicity and to use a transform in which wavelets can be oriented in any direction. This makes for a special kind of "mathematical microscope." As J. C. van den Berg writes in *Wavelets in Physics*, "a real-world microscope is not more sensitive in one direction than in another one.... But the mathematical microscope as embodied in two-dimensional wavelets has an extra feature: these wavelets can be designed in such a way that they are *directionally selective*. Apart from dilation and translation, one can now also *rotate* the wavelet."

Three different techniques have been devised to provide this flexibility: a continuous two-dimensional transform, a discrete oversampled transform that uses "steerable" filters, and special waveforms called *brushlets*, which combine steerability and orthogonality.

The Continuous Two-Dimensional Wavelet Transform

The continuous two-dimensional wavelet transform was designed in 1987 by Romain Murenzi (see [Murenzi] and [Antoine et al. 2]). The key to its design was the fact, known since 1972, that coherent states (invented by Schrödinger in 1926 and then used in the 1960s to describe coherent light, such as lasers) can be derived from group representations. In 1985, Alex Grossmann, Jean Morlet, and Thierry Paul discovered that the one-dimensional wavelets used in the continuous wavelet transform are the coherent states associated with the affine group of the line. Murenzi applied this insight to the design of the continuous two-dimensional wavelet transform when he was working on his Ph.D. at the Catholic University of Louvain in Belgium, under the direction of Jean-Pierre Antoine and Grossmann. Applications include measuring the velocity field of a turbulent flow around an obstacle, fault detection in geology, and uncovering hidden symmetries in two-dimensional patterns.

Up until now, this continuous transform has been computationally expensive. But building upon an idea of Bruno Torrésani [Torrésani], two students of Antoine's, Pierre Vandergheynst and Jean-François Gobbers, recently developed a discrete version of the two-dimensional continuous transform, which can be used with wavelets that are highly sensitive to directions. Compared to continuous transforms based on the fast Fourier transform, this fast algorithm cuts the number of computations by a factor roughly equal to the logarithm of the size of the signal, typically of the order of 10 for a one-dimensional signal, and more in the two-dimensional case.

Steerable Filters

Steerable filters provide a different approach to the problem of orientation (see also the discussion of steerable filters in *Wavelets and Vision: Another Perspective*, p. 231, and Eero Simoncelli's web page on steerable pyramids: http://www.cis.upenn.edu/˜eero/steerpyr.html). These filters, developed by Adelson and William T. Freeman, make it possible to determine the response of a filter of any orientation without explicitly applying that filter.

You begin, Adelson says, with a finite set of oriented filters, for example, three filters, each producing a subband at a different orientation. "Given these three filters you can calculate the response of a new filter rotated to an arbitrary angle by taking a linear combination of those three. Thus the filters are a linear basis for the space of rotated versions of themselves. This means two things. First, you know that you are treating all orientations uniformly. Second, you can compute a small set of subbands, and then synthesize any subband you desire without doing any extra filtering. You simply sum the basis subbands, and you get the subband you would have gotten had you applied a filter at the desired orientation."

Steerable filters need not be wavelets, but in practice, Adelson says, they are almost always wavelet-like. Those he uses "do not fit into the classic wavelet math, but they are localized in space, frequency, and orientation as wavelets are."

"I continue to be puzzled about how to use the term 'wavelet,'" he adds. "One can restrict it to a precise mathematical class, or one can speak of filters that have general wavelet-like qualities. In either case one invites confusion."

Designing steerable filters is easy, Adelson says, but building a good "pyramid" from them for use in fast computations is difficult. In addition to translation invariance and self similarity, one would like the analysis filters used to build the pyramid to be the same as the synthesis filters used to reconstruct, and the pyramid should be efficiently subsampled. "Unfortunately it is impossible to satisfy all these goals simultaneously," Adelson says. "Steerable pyramids are engineering compromises that satisfy several goals as well as possible. Thus they lack the mathematical elegance of an orthogonal wavelet transform, but they have much more utility for image processing and machine vision."

Brushlets: Painting with "Wavelets"

A different approach is taken by Ronald Coifman and François Meyer, Yves Meyer's son, who came to Yale University from France in 1993 and now works on the analysis of the motion of the heart from MRI images, motion estimation, video compression, and image compression.

The steerable filters described above are significantly overcomplete, which is a drawback for image compression. In addition, they are not orthogonal, so there is no efficient algorithm for adaptively selecting the set of filters that can describe an image most concisely.

In order to combine angular resolution and an adaptive algorithm that finds the most concise representation, Coifman and Meyer devised a way to "paint" an image using a collection of "brush strokes" or "brushlets" [Meyer, Coifman] (Figure 1). A brushlet is an oriented filter that can be tuned to any single frequency. The library of brushlets, with all possible frequencies, sizes, and locations, constitute an overcomplete representation, with more "building blocks" than necessary, as in the case for traditional oriented filters. But it is possible to select from the brushlet library an optimal orthogonal decomposition, using the fast Best Basis algorithm (see pages 96 and 259).

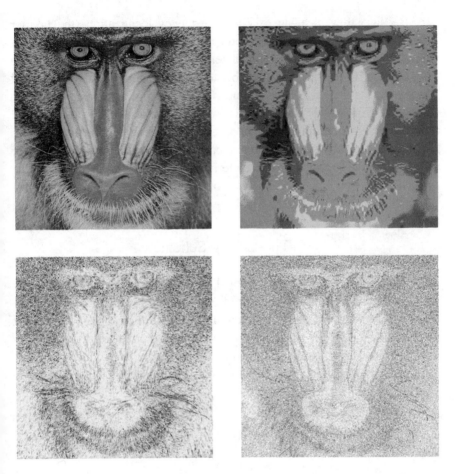

Figure 1. This figure illustrates the fact that different mathematical bases provide different *styles* for representing images. The style of a given basis corresponds to the set of features that are efficiently represented with such a basis.
We can decompose an image as a sum of layers, where each layer is "painted" with a different style. The sum of the layers gives the original image back. Here we illustrate this decomposition with three painting styles (i.e., three different mathematical bases): 1. watercolor: wavelets (top right); 2. Van Gogh style: brushlets (bottom left); 3. Pointillist "à la Seurat": Diracs (bottom right). The original image is the sum (in the mathematical sense) of the three images: watercolor, Van Gogh, and Seurat (top left). For each layer, only a very small number of coefficients in each basis were required to represent the layer. (Courtesy of François Meyer.)

Because they have an excellent frequency localization (the Fourier transform of the brushlet has a compact support, while the brushlet itself is well localized in the spatial, or temporal, domain), brushlets provide very sparse representations of highly textured images and have been used successfully for image compression; it's possible, as Meyer puts it, for the "brushlet painter" to "paint with a few brush strokes the texture in a field of wheat." In addition, he says, because of its mathematical definition, "the brushlet basis is an excellent basis for the numerical solutions of acoustic and electromagnetic scattering problems." (A brushlet is the Malvar wavelet transform of the Fourier transform of a function; for Malvar wavelets, see p. 93.)

The Lifting Scheme

A third approach to two-dimensional wavelet transforms is provided by the *lifting scheme*. For a number of applications in two dimensions—in geophysics and geography, for example—data is more naturally thought of as existing on the surface of a sphere, rather than on a plane. Peter Schröder and Wim Sweldens have shown how to produce spherical wavelets for use in such applications (see [Schröder, Sweldens] and *The Lifting Scheme*, p. 257).

Pyramid Algorithms of Burt and Adelson

Peter Burt and Edward Adelson began working together while they were post-doctoral fellows living in Manhattan, with Adelson working at New York University and Burt commuting to the University of Maryland. "We were both intrigued by the multiscale ideas around at the time, like those of David Marr at M.I.T., and we had independently worked out the idea of constructing an iterative multiscale decomposition," Adelson says. "Our initial work was based on nothing more complicated than intuitions and basic algebra." Burt had introduced pyramid algorithms for machine vision; in a joint publication he and Adelson introduced the terms "Gaussian" and "Laplacian" pyramids, and showed how they could be used for image compression [Adelson, Burt]. The better-known Burt and Adelson paper on pyramid algorithms [Burt, Adelson] was published in 1983. "I suspect," one reviewer wrote, "that no one will ever use this algorithm again." Then in 1986, working with Eero Simoncelli (then a summer student at RCA Labs, now at M.I.T.), Adelson constructed orthogonal pyramids [Adelson et al.], "and later learned that we had re-invented quadrature mirror filters." In 1989 Adelson and Simoncelli received a patent for the use of the transform in image data compression. "The patent is held by GE—having acquired RCA—and may well cover most wavelet-based image coding," Adelson says. "But I don't know if GE is aware of it." (Others question to what extent the patent could apply to such encoding.)

Multiwavelets

Daubechies created the first orthogonal wavelets with compact support using iterations: a Daubechies wavelet is the limit of an iterative process and cannot be created from analytic formulas. More recently other researchers have found that one *can* make orthogonal wavelets with compact support that are explicit functions, by using more than one scaling function.

The resulting multiresolution analysis, with multiple scaling functions and wavelets, is not subject to the limitations of Daubechies wavelets. It is possible, for example, to create symmetrical orthogonal wavelets with compact support. (Symmetry is sometimes considered desirable in image encoding; see p. 243. The only symmetrical Daubechies wavelet is the discontinuous Haar wavelet.) "The objective is to use multiwavelets to construct wavelets with varying degrees of regularity, short support, symmetry, orthogonality ...," write George Donovan and Jeffrey S. Geronimo of the Georgia Institute of Technology and Douglas P. Hardin of Vanderbilt University [Donovan et al. 1].

"For instance," says Geronimo, "if you use two scaling functions you can get symmetry. If you have three scaling functions then you can construct piecewise linear wavelets, with closed-form formulas. We can also construct wavelets that are piecewise linear and symmetric or antisymmetric, but for that you need four scaling functions."

He and his colleagues have found two procedures for creating multiwavelets. The first produces "fractal" wavelets. They start with a wavelet that generates a non-orthogonal multiresolution analysis, such as the "hat" function, which equals $1 - |x|$ for $-1 \leq x \leq 1$ and 0 elsewhere. They add to it (point by point) a *fractal interpolation function,* chosen so that when the part of the new function (the hat function plus the fractal interpolation function) on [–1,0] is shifted to [0,1], it is orthogonal to the part on [0,1]. (For a discussion of fractal interpolation functions, see [Barnsley].)

This procedure gives rise to two symmetric, orthogonal, compactly supported, continuous scaling functions. They are then used to construct two wavelets, one symmetric and other antisymmetric. While in the multiresolution analysis of Mallat and Meyer we have a single "mother" wavelet, all of whose translates and dilates are orthogonal to each other, here we have two "mother" wavelets, all of whose translates and dilates are orthogonal to each other. (P. R. Massopust of Sam Houston State University has also worked on this; see [Donovan et al. 3].) The work with fractal wavelets has been extended to two dimensions.

The second procedure produces piecewise polynomial wavelets (see [Donovan et al. 1], [Donovan et al. 2]). Wavelets created this way are at least continuous, and some have been created that are once continuously differentiable. (*Piecewise polynomials* are functions whose graphs are made by stringing together pieces of graphs of polynomials. First-degree piecewise polynomials are made of straight lines, whereas second-degree piecewise polynomials are composed of arcs of parabolas. Piecewise polynomial wavelets were among the first orthogonal wavelets created, but those—the Battle-Lemarié spline functions—did not have compact support.)

Is there a price to pay? "Since there are more scaling functions at first sight one might think there will be a computational cost," says Geronimo. But multiwavelets satisfy the same deconstruction and reconstruction algorithms as single wavelets, and the short support of the scaling functions may make them computationally efficient. "This area is still very young, and as far as I know there has not yet been much comparison computationally between multiwavelets and single wavelets."

The Heisenberg Uncertainty Principle and Time-Frequency Decompositions

It is sometimes said that one can't localize a signal simultaneously in time and frequency. This can be misleading. It's not that *we* can't localize the signal in time and frequency. The *signal* itself can't be concentrated simultaneously in time and in frequency. The Heisenberg uncertainty principle does not limit what we can know about reality; it describes that reality. The shorter-lived a function, the wider the band of frequencies given by its Fourier transform; the narrower the band of frequencies of its Fourier transform, the more the function is spread out in time.

More precisely, the Heisenberg principle says the following (a proof is in the Appendix, p. 281). For every function $f(t)$ (t a real number), such that

$$\int_{-\infty}^{\infty} |f(t)|^2 \, dt = 1, \tag{12.1}$$

the product of the variance of t and the variance of τ (the variable of $\hat{f}$) is at least $\frac{1}{16\pi^2}$:

$$\underbrace{\left(\int_{-\infty}^{\infty} (t - t_m)^2 |f(t)|^2 dt\right)}_{\text{variance of } t} \underbrace{\left(\int_{-\infty}^{\infty} (\tau - \tau_m)^2 |\hat{f}(\tau)|^2 d\tau\right)}_{\text{variance of } \tau} \geq \frac{1}{16\pi^2}$$

$$\tag{12.2}$$

(The exact number at the right depends on the formula used for the Fourier transform; see p. 275. The normalizing condition (12.1) is reasonable since in quantum mechanics this integral measures probability, and probability 1 describes an event that is certain.)

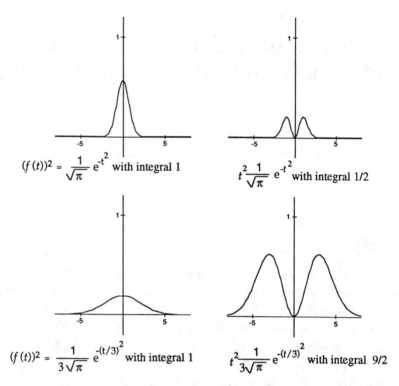

Figure 1. Left: two functions. Right: the same functions multiplied by t^2 (i.e., $(t - t_m)^2$ where $t_m = 0$). This multiplication emphasizes the values of the functions for big values of t and minimizes them for small values of t. The integrals at right measure the *variance* of t for the corresponding functions at left. The function at upper left is concentrated in time, and has a small variance; the function at lower left is spread out in time and has a larger variance.

These *variances* measure to what extent t and τ take values far from their average values, t_m and τ_m. (The notion of variance is discussed in *Probability, Heisenberg, and Quantum Mechanics*, p. 211.) Thus the more f is concentrated in a small window of time, the smaller the variance of t will be; if the signal is more spread out in time, the variance will be larger. (Multiplying $|f(t)|^2$ by $(t - t_m)^2$ emphasizes the values of $|f(t)|^2$ for big values of $|t - t_m|$—that is, when t is far from its average—and minimizes them when $|t - t_m|$ is small.)

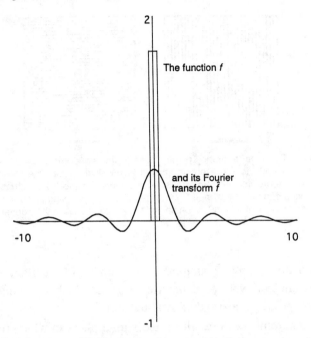

Figure 2. A function and its Fourier transform cannot both be very concentrated.

Some examples illustrate this. On the left of Figure 1 we show two functions and their integrals; on the right we have the same functions multiplied by t^2 (i.e., $(t - t_m)^2$ where $t_m = 0$).

The second integral in (12.2) measures the range of frequencies of the Fourier transform of f. The smaller the variance of τ, the narrower the band of frequencies of $\hat{f}$.

Now let's consider a function $|f(t)|^2$ whose graph is very concentrated around $t = 0$. Since we normalized the integral of this function to be 1 (condition (12.1)), the function looks like a peak—and the Fourier transform of a peak is necessarily very spread out, as shown in Figure 2.

Time-Frequency Representations

Whatever method you use to decompose a signal simultaneously into time and frequency, you'll run into this problem. If you want precise

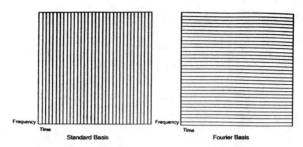

Figure 3. The time-frequency plane. Left: in the standard or Dirac "basis"— which corresponds to representing signals in physical space—the Heisenberg boxes degenerate into infinite vertical lines. Right: in the Fourier "basis" the boxes become infinite horizontal lines. In practice, the sampling interval leads to tall, skinny boxes in the standard basis and short, wide boxes in the Fourier basis.

information about time, you have to put up with a certain vagueness about frequency; if you want precise information about frequency, you have to accept a certain vagueness about time.

This compromise can be illustrated with a plane in which time varies horizontally and frequency vertically (an idealized version of the time-frequency plane, as mentioned in *Best Basis*, p. 259). This plane can be tiled with rectangles of size Δt by $\Delta \tau$. (As mentioned in *Best Basis*, Heisenberg boxes do not tile the time-frequency plane exactly; they

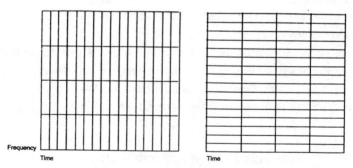

Figure 4. An idealized time-frequency plane decomposed with windowed Fourier analysis. Left: a narrow window, which gives more precision about time, less about frequency. Right: a wide window, giving more precision about frequency, less about time.

overlap.) "Delta t" represents the window of time; it is the standard deviation of time—the square root of the variance of t in equation (12.2). "Delta tau" represents the range of frequencies; it is the standard deviation of frequency—the square root of the variance of τ in equation (12.2). Victor Wickerhauser calls these rectangles *Heisenberg cells* or *Heisenberg boxes*.

The Heisenberg uncertainty principle says that the area of each cell, or box, must be at least $1/4\pi$, but depending on the basis used, the cells used will have different shapes and positions. "For example, in the standard or 'Dirac' basis, which pinpoints the instants when events occur, the plane is tiled with the tallest, thinnest patches allowed by the sampling interval," Wickerhauser says. The Fourier basis pinpoints the frequencies present in events that occur over long periods, so the Heisenberg boxes are all short (good information about frequency) but they are very wide, as shown in Figure 3.[1]

With windowed Fourier analysis, the form of the Heisenberg boxes depends on the size of the window, as shown in Figure 4. A small window looks at short intervals of time at the cost of being vague about frequency. A big window is less precise about time but more precise about frequency. In both cases, the size of the window—the width of the box—remains fixed for each decomposition: the same for high frequencies and for low frequencies.

With wavelets, a wide window is used for low frequencies and narrower and narrower windows for high frequencies, as illustrated by Figure 5. But, as required by Heisenberg, the Heisenberg boxes that correspond to narrow wavelets are taller: time is privileged at the expense of frequency.

"This kind of analysis of course works best if the signal is composed of high-frequency components of short duration plus low-frequency components of long duration, which is often the case with signals encountered

---

[1]Strictly speaking, these are not bases. The Dirac "functions" of the standard "basis" are not really functions, and cannot be elements of any basis. The notion of "Fourier basis" is correct for Fourier series; the functions $\sin 2\pi kt$ and $\cos 2\pi kt$ are elements of $L^2[0, 1]$. But the "basis functions" $e^{2\pi i \tau t}$ for the Fourier transform are not elements of $L^2(\mathbb{R})$; they are not square integrable.

208

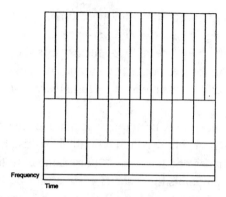

Figure 5. An idealized time-frequency plane decomposed with wavelets. The size of the window varies. The wide windows used to look at low frequencies are precise about frequency, vague about time. The narrow windows used to look at high frequencies are precise about time, vague about frequency.

in practice," write Olivier Rioul of the Ecole Nationale Supérieure des Télécommunications in France and Martin Vetterli of the University of California at Berkeley and the Swiss Federal Institute of Technology in Lausanne [Rioul, Vetterli, p. 18].

Probability, Heisenberg, and Quantum Mechanics

—"...one can't believe impossible things." "I daresay you haven't had much practice," said the Queen. "When I was your age, I always did it for half-an-hour a day. Why, sometimes I've believed as many as six impossible things before breakfast."

—Lewis Carroll, *Through the Looking-Glass*

It is not extremely difficult to accept the fact that a signal can't be "localized" simultaneously in time and in frequency. Speaking of a frequency at a given instant doesn't really have a meaning; a frequency has to have the time to oscillate. It is much harder to come to terms with the fact that one cannot speak of an elementary particle's precise position *and* precise momentum. One may be tempted to think that Heisenberg imposes a limit on what one can *know* (lacking the necessary intelligence or tools). That is not the case. As in signal processing, the Heisenberg principle describes a reality: an elementary particle does not have simultaneously a precise position and a precise momentum.

But in order even to state the uncertainty principle in the context of quantum mechanics we will need to introduce some new vocabulary. In discussing the position and momentum of elementary particles, two issues are involved. First (and this is independent of quantum mechanics), any such discussion is necessarily probabilistic. We don't have the right to speak about an elementary particle's position. The question that has meaning is: what is the probability that the particle is in a given region? Nor can we speak of a particle's momentum. The question that has meaning is: what is the probability that the momentum is between two given values? (This does not mean that we can't know the position precisely, or the momentum precisely: there is no lower bound to how small the region of position can be, or how small the interval of momentum.)

Second, quantum mechanics tells us that certain pairs exist such that (as for time and frequency) more information about one means less information about another: position and momentum form one such pair; the x and y components of spin form another.

While the inherent strangeness of quantum mechanics is not related to probability theory—statistical mechanics, which relies on probabilities, is a subject of classical physics—the language of probability is needed for any discussion of quantum mechanics, so we will introduce it first.

The Language of Probability

Probability only became a respectable part of mathematics in the 1930s, with the discovery by the Russian mathematician Andrei Kolmogorov that the theory of probability and measure theory are the same thing. This made it possible to give a precise meaning to notions that until then had been somewhat vague.

A *probability space* X is a set (typically, the set of possible outcomes of an experiment) along with a function P that measures subsets of X. We throw two dice. Since the two dice each have six sides, with values from 1 to 6, we say that the set X has 36 *elements* (in the language of set theory) or that the experiment has 36 *outcomes* (in the language of probability theory). If the dice aren't loaded, each possible outcome (or element of X) has probability 1/36. The *event* or subset "the total is 5" has probability $4/36 = 1/9$. (An event is thus composed of different outcomes.) We can also say that P gives to the subset $\{(1,4),(2,3),(3,2),(4,1)\}$ the *measure* 1/9.

A *random function* (often called a *random variable*) is a function $X \rightarrow \mathbb{R}$ (that is, a function that assigns a real number to each outcome). One should think of a quantity that can be measured each time one does the experiment; for example, the total value obtained when two dice are thrown.

The principal construction in probability theory is the *mean* of a random function, $M(f)$, also called its *expectation*. In the case of our dice, the mean of "total value" is

$$2\frac{1}{36} + 3\frac{2}{36} + 4\frac{3}{36} + 5\frac{4}{36} + 6\frac{5}{36} + 7\frac{6}{36} + 8\frac{5}{36} + 9\frac{4}{36}$$
$$+10\frac{3}{36} + 11\frac{2}{36} + 12\frac{1}{36} = 7.$$

(The first term represents the only case, of the 36 possible outcomes, where the total value of the two dice is 2; the second term, the two ways one can get a total of 3, etc.)

The *standard deviation* of a random function f, denoted $\sigma(f)$ ("sigma" f), measures how spread out the values of f are around its mean. The average deviation from the mean, $f - M(f)$, is zero: it is as often negative as positive. To make the standard deviation positive, first we square the average deviation, then we take the square root of the result:

$$\sigma(f) = \sqrt{M(f - M(f))^2}$$

Taking the square root maintains the right units. The average of the squares of deviations

$$\text{Var}\,(f) = M((f - M(f))^2),$$

called the *variance* of f, cannot be considered an average deviation because it has the wrong units. If, for example, we measure the height in centimeters of children of a certain age, the random function is a length, whereas the variance is measured in cm^2. The standard deviation—the square root of the variance—does have the right units.

In the above example, where T represents the total obtained when throwing two dice, we have

$$\text{Var}\,(T) = \frac{1}{36}(2 - 7)^2 + \frac{2}{36}(3 - 7)^2 + \cdots + \frac{6}{36}(7 - 7)^2 + \cdots +$$
$$+\frac{2}{36}(11 - 7)^2 + \frac{1}{36}(12 - 7)^2 = 5.833....$$

The standard deviation in this example is thus $\sqrt{5.833\ldots} \approx 2.415\ldots$

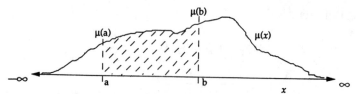

Figure 1. When probability spaces are continuous, the probabilities of events cannot be described in terms of individual outcomes; integrals are needed. The function μ is a probability density. The real numbers between $-\infty$ and ∞ represent all the theoretically possible outcomes of an experiment, while the integral $\int_a^b \mu(x)\,dx$—the shaded zone—measures the probability of an outcome between a and b.

Probability Expressed as Integrals

As long as a probability space is discrete, all problems can be understood in terms of the probabilities of the individual outcomes. The outcomes are then countable (or even finite, as in the case of our dice), and although the language of measure theory is useful, it is not essential. Serious problems in probability involve probability spaces that are not just infinite but *continuous*; in this case, the probabilities of events cannot be described in terms of the probabilities of individual outcomes, and the language of measure theory is essential.

Here is a typical situation. Suppose that we have a function μ ("mu"), whose value is never negative, that delimits between $-\infty$ and ∞ an area equal to 1, as shown in Figure 1.

The function μ is a probability density. The real numbers between $-\infty$ and ∞ represent all the theoretically possible outcomes of an experiment, while the integral

$$P[a, b] = \int_a^b \mu(x)\,dx,$$

represented in the figure by the area of the shaded zone, measures the probability of getting an outcome between a and b. The integral

$$P[-\infty, \infty] = \int_{-\infty}^{\infty} \mu(x)\,dx = 1 \tag{13.1}$$

measures the probability, which is certain, that one of the theoretically possible outcomes will actually occur.

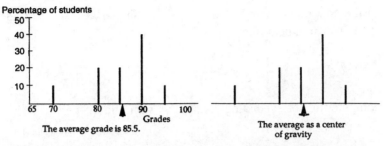

Figure 2. The mean of a random function $f(x) = x$ can be thought of as the center of gravity of a bar of density $\mu(x)$. Here the grades (x) obtained by 10 students are weighted by the fraction of students receiving each grade ($\mu(x)$), and are represented as bars on a scale. The average grade is its center of gravity.

Note that the probability of getting the exact outcome a is zero; there is no "space" under the function μ at the point a. (That an infinite number of zeros add up to 1 may seem bizarre, but it's the same as saying that a line has a length, while the points that compose it have length zero—an idea that everyone is used to accepting.) It is impossible, then, to understand this continuous probability space by calculating the probabilities of individual outcomes. One must replace these sums (finite or infinite) by integrals.

When a probability space is continuous, the mean of the random function f is given by

$$M(f) = \int_{-\infty}^{\infty} f(x)\mu(x)\, dx. \tag{13.2}$$

In the case where $f(x) = x$ (which we can call simply the function x), we can think of this mean as the "center of gravity" of a bar of density $\mu(x)$. This image will be useful when we talk about quantum mechanics; to illustrate it we will return for a moment to a discrete probability space.

Ten students take an exam. One gets the grade 95, four get 90, two get 85, two get 80, and one gets 70, as shown in Figure 2. Each possible grade x is given the weight $\mu(x)$. For the grade 95, that weight is $\frac{1}{10}$, since one student out of ten gets 95; for the grade 90, the weight is $\frac{4}{10}$. To compute the average grade, we multiply each grade by its weight $\mu(x)$ and add the results.

In the continuous case, we do the same thing. But instead of adding a sequence of numbers like

$$95\left(\frac{1}{10}\right) + 90\left(\frac{4}{10}\right) + 85\left(\frac{2}{10}\right) + 80\left(\frac{2}{10}\right) + 70\left(\frac{1}{10}\right) = 85.5,$$

we compute the corresponding continuous sum, the integral

$$M(x) = \int_{-\infty}^{\infty} x\mu(x)\,dx. \tag{13.3}$$

When scientists study the behavior of turbulent flows, or engineers work to improve the internal combustion engine, their primary mathematical tool is thus the integral. They cannot hope to deal with one molecule at a time; any picture they get of reality at a macroscopic level is necessarily based on statistical (essentially, probabilistic) knowledge of what is going on at a microscopic level.

Quantum Mechanics

In classical mechanics, position and momentum are functions of the state of a system. Throw a tennis ball in an empty room: the room and the ball make up a system, each state of the system consisting of a position and a momentum for the ball. We can observe and measure these quantities; we can also use known laws to predict their evolution. If instead of a tennis ball we have a chamber filled with gas, the problem is of course more involved, and we will come up with some statistical description of what is going on. This is complicated by the fact that the act of measuring position changes momentum and vice versa. But this latter complication is not inherently strange. If it is not actually true that the watched pot never boils, the proverb reflects a common reality: if an elementary school principal sits in on a class to observe how well a novice teacher maintains discipline, the behavior of the children will inevitably be affected by the "act of measurement."

Quantum mechanics is different, giving a picture of reality at odds with our experience of the world. In quantum mechanics, the state of a system is represented by the wave function, $\psi(x)$, which evolves according to Schrödinger's equation (the ψ representing the wave function not, of course, to be confused with the ψ representing a wavelet). If this perfectly deterministic equation seems at first glance the formula Laplace dreamed of, which "would encompass... all the movements of the largest bodies of the universe and those of the lightest atom," it is not true that it reveals to us "the future, like the past."[1]

We can measure a particle's position or momentum (one at a time, the act of measuring position changing momentum and vice versa), and we can use the wave function to compute the probability that particles will be in a certain region, or have a certain momentum. But the Heisenberg uncertainty principle tells us that an elementary particle doesn't simultaneously *have* a precise position and precise momentum, even one measured in terms of probabilities.

We will describe with a little more precision this complicated situation, in particular showing how the Heisenberg principle applies. To simplify matters, we will stick to one-dimensional physics, limiting ourselves to a single particle on a straight line with coordinate x (where x is a real number).

As we said, a state of this system—its complete description at a given instant—is represented by the wave function, $\psi(x)$. This function has a phase, which makes it extremely difficult to understand. So temporarily let us cheat and think instead of the function $|\psi(x)|^2$, which is easier to comprehend; we can think of it as a "density of probability of presence." (It replaces $\mu(x)$ in formulas (13.1), (13.2), and (13.3).)

---

[1] Actually, if one can believe quantum mechanics, the wave function does predict the future, but the state of the world that the wave function encodes is very perplexing, as exemplified by Schrödinger's cat. Schrödinger imagined a cat put in a box with a radioactive material with a probability 1/2 that one particle would decay in the next hour, triggering release of a poison. After an hour, will the cat be alive or dead? The highly unsatisfactory answer of the wave function is that it will be some superposition of live and dead cats. Quantum mechanics does not say that the cat is alive or dead and you don't know which; it says that the superposition of live and dead cats *is* the reality. Of course when you open the box, the cat will either be alive or dead; this means that the wave packet has been "reduced".

The random function for the position of our particle is x (remember we are in one dimension). The probability that the particle is somewhere on the line is given by the integral[2]

$$P[-\infty, \infty] = \int_{-\infty}^{\infty} |\psi(x)|^2\, dx = 1.$$

The probability that the particle is between a and b is given by the integral

$$\int_{a}^{b} |\psi(x)|^2\, dx. \tag{13.4}$$

Its average position (x_m, the mean of the function x) is obtained by integrating the product of x and the density of probability $|\psi(x)|^2|$:

$$x_m = \int_{-\infty}^{\infty} x|\psi(x)|^2\, dx. \tag{13.5}$$

This formula corresponds to formula (13.3) above. Computing the average position of our particle is the continuous equivalent of our discrete example of computing the average grade on an exam, or the center of gravity of a rod.

The variance of x is

$$\mathrm{Var}\,(x, \mu_\psi) = \int_{-\infty}^{\infty} (x - x_m)^2 |\psi(x)|^2\, dx.$$

What about momentum? We said earlier that we would cheat temporarily and not think of the wave function $\psi(x)$, whose phase makes it difficult to understand, but rather of the easier $|\psi(x)|^2$. But we can't think of the density of probability for momentum in terms of $|\psi(x)|^2$. There is no random function for momentum in the "position" side of the Fourier transform. To find one, we must cross over the Fourier transform

---

[2] We see that $\psi(x)$ is *square summable*: the integral of its absolute value squared is finite. The wave function thus lives in the function space L^2, discussed in *Traveling from One Function Space to Another*, p. 223. Since this integral equals 1, a state of our system is a *normalized* element of $L^2(\mathbb{R})$.

to what we called the "momentum representation." The random function for momentum is $h\xi$, where h is Planck's constant, with the value $6.624 \times 10^{-27}\,\text{g cm}^2/\text{s}$ (grams × centimeters squared, per second).[3]

The probability that the momentum of our particle is between the values $h\alpha$ and $h\beta$ is

$$\int_{\alpha}^{\beta} |\hat{\psi}(\xi)|^2 d\xi. \tag{13.6}$$

The mean value of the particle's momentum, which we will call $h\xi_m$, is given by

$$h\xi_m = \int_{-\infty}^{\infty} h\xi |\hat{\psi}(\xi)|^2 \, d\xi,$$

and its variance by

$$\text{Var}(h\xi, \mu_\psi) = \int_{-\infty}^{\infty} (h\xi - h\xi_m)^2 |\hat{\psi}(\xi)|^2 \, d\xi.$$

The Uncertainty Principle

Now we are ready to return to Heisenberg's uncertainty principle. In *The Heisenberg Uncertainty Principle and Time-Frequency Decompositions* we gave this principle in the following form:

$$\underbrace{\left(\int_{-\infty}^{\infty} (t - t_m)^2 |f(t)|^2 dt\right)}_{\text{variance of } t} \underbrace{\left(\int_{-\infty}^{\infty} (\tau - \tau_m)^2 |\hat{f}(\tau)|^2 d\tau\right)}_{\text{variance of } \tau} \geq \left(\frac{1}{4\pi}\right)^2.$$

We can now rewrite it in terms of quantum mechanics:

$$\underbrace{\left(\int_{-\infty}^{\infty} (x - x_m)^2 |\psi(x)|^2 dx\right)}_{\text{variance of position}} \underbrace{\left(\int_{-\infty}^{\infty} (h\xi - h\xi_m)^2 |\hat{\psi}(\xi)|^2 d\xi\right)}_{\text{variance of momentum}} \geq \left(\frac{h}{4\pi}\right)^2.$$

---

[3] Aside from the fact that h is necessary to describe physical reality, this gives us the right units. The wave number (or spatial frequency) ξ is measured in 1/cm. Momentum is measured in mass × velocity. Multiplying ξ by Planck's constant gives us those units: $\frac{\text{gcm}^2}{\text{s}} \times \frac{1}{\text{cm}} = \frac{\text{gcm}}{\text{s}}$.

If we keep clearly in mind the picture of a function and its Fourier transform, the transform becoming more and more spread out as one compresses the function, and vice versa, we are forced to admit that, bizarre as it may appear, the uncertainty principle describes physical reality and not (as one might be tempted to think) a gap, even irremediable, in our knowledge of that reality.

Quantum Mechanics, Operators and the Quabability

Physicists would like to be able to work with position, momentum, and other quantities without constantly changing the form of representation. The state of the physical world is represented by the wave function, $\psi(x)$. Can we study that function without straddling the Fourier transform, using one representation of ψ for position, and another for momentum?

Quantum mechanics gives the following answer: the wave function is a vector in a Hilbert space. Just as in signal processing we can represent a signal (i.e., a vector in an infinite-dimensional space) in physical space, or as its Fourier transform, or as a wavelet transform (using any number of different wavelet bases) we can represent the wave function in any one of many different bases in the Hilbert space. (The "position representation" is one such basis, the "momentum representation" another; both "identify" the abstract Hilbert space of quantum mechanics with $L^2(\mathbb{R})$, the first when the real numbers $\mathbb{R}$ describe position, and the second when they describe momentum.)

The very term "wave function," hinting at the more familiar and concrete sound waves or electromagnetic radiation, suggests that we are thinking of ψ in the position basis. So let us adopt a more abstract term, *quabability*, a word combining "quantum mechanics" and "probability."[4] The state of the physical world is represented by the quabability Q, which lives in the abstract vector space known as Hilbert space. The quabability has as many manifestations as there are bases in the Hilbert space; in the position basis it appears as ψ; in the momentum basis it appears as $\hat{\psi}$.

---

[4]To the best of our knowledge, the word *quabability* was coined by Jeremy Bem while he was an undergraduate at Cornell University.

The quabability encodes probabilities concerning various aspects of the physical world (often called *observables*)—position and momentum are two, but there are others: angular momentum, spin, etc. It does this through *operators*: the scalar product of the quabability and the position operator acting on the quabability gives the expectation (or mean) for position, the scalar product of the quabability and the momentum operator acting on the quabability gives the expectation for momentum, and so on.

This is all quite abstract, and if you wanted to actually compute any probabilities concerning position, momentum, or spin, the above statement would not give the vaguest idea where to begin. Operators in quantum mechanics are rules that allow us to compute probabilities, but each operator has as many manifestations as there are bases in the Hilbert space. There is no single answer to the question: "How does the position operator compute probabilities?" The right question is: "How does the position operator act on the quabability to compute position probabilities *in such-and-such a basis*?" Similarly, the right question to ask about the momentum operator is: "How does it act on the quabability to compute momentum probabilities *in such-and-such a basis*?"

So if we want to study probabilities for position and momentum of a particle while remaining in the position basis, we need to ask: "How do the position and momentum operators work in the position basis?"

Traditionally, for reasons not at all evident, the position operator is denoted q and the momentum operator is denoted p—even worse, just asking to be confused with position. As we have trouble keeping that notation straight, we will denote the position operator O_{pos} and the momentum operator O_{mom}.

In the "position basis," the position operator O_{pos} is simple. Theory and experiments agree that in the position basis, the position operator multiplies the quabability by the function x:

$$\underbrace{O_{pos}Q}_{\substack{\text{position operator} \\ \text{in Hilbert space}}} \quad \text{becomes} \quad \underbrace{(O_{pos}(\psi))(x) = x\,\psi(x)}_{\substack{\text{what position operator} \\ \text{does in position basis}}}.$$

(We use ψ rather than Q to denote the quabability when it is in the position basis.)

The mean position is then given by the scalar product

$$\underbrace{\langle Q, O_{pos}Q \rangle}_{\substack{\text{mean position} \\ \text{in Hilbert space}}} = \underbrace{\langle \psi, x\psi(x) \rangle}_{\substack{\text{mean position} \\ \text{in position basis}}} = \underbrace{\int_{-\infty}^{\infty} x|\psi(x)|^2 \, dx}_{\text{same as Equation 13.5}}.$$

But if we cross the Fourier transform to go into the momentum basis the position operator no longer multiplies the wave function ψ by x; instead it differentiates $\hat{\psi}$:

$$\underbrace{O_{pos}(Q)}_{\substack{\text{position operator} \\ \text{in Hilbert space}}} \quad \text{becomes} \quad \underbrace{(O_{pos}(\hat{\psi}))(\xi) = \frac{1}{2\pi i}\frac{d\hat{\psi}}{d\xi}}_{\substack{\text{what position operator} \\ \text{does in momentum basis}}}.$$

In contrast, in the momentum basis, the momentum operator corresponds to multiplication of $\hat{\psi}$ by the function $h\xi$:

$$\underbrace{O_{mom}(Q)}_{\substack{\text{momentum operator} \\ \text{in Hilbert space}}} \quad \text{becomes} \quad \underbrace{(O_{mom}(\hat{\psi}))(h\xi) = h\xi\hat{\psi}(\xi)}_{\substack{\text{what momentum operator} \\ \text{does in momentum basis}}}.$$

(We use $\hat{\psi}$ rather than Q to denote the quability in the momentum basis.) But in the position basis that multiplication becomes differentiation:

$$\underbrace{O_{mom}(Q)}_{\substack{\text{momentum operator} \\ \text{in Hilbert space}}} \quad \text{becomes} \quad \underbrace{(O_{mom}(\psi))(x) = -i\hbar\frac{d\psi}{dx}}_{\substack{\text{what momentum operator} \\ \text{does in position basis}}}.$$

The mean momentum is then the scalar product:

$$\underbrace{\langle Q, O_{mom}Q \rangle}_{\substack{\text{mean momentum} \\ \text{in Hilbert space}}} = \underbrace{\langle \psi, -i\hbar\frac{d\psi}{dx} \rangle}_{\substack{\text{mean momentum} \\ \text{in position space}}}.$$

This creates problems. If you were to toss a coin with your left hand while throwing a die with your right, you could compute the probability of getting heads and a 6 by multiplying $\frac{1}{2}$ by $\frac{1}{6}$, in either order. But differentiation and multiplication *do not commute*: if you multiply the two operators together to compute the probability of a particle being simultaneously in some interval with a certain range of momentum, your answer will depend on the order in which you do the multiplication. To avoid that problem, we need to use the two sides of the Fourier transform— landing us back on the lower bound imposed by the uncertainty principle. In position space there exist states of our system in which the probability is extremely high—even 1—that the particle is in as small an interval as one likes. In momentum space, states exist in which the probability is extremely high—even 1—that the particle's momentum is in as small a range of values as one wishes. But these states are not the same.

Note that while mathematically, either things commute or they don't, in practice, on a human scale, the two operators *almost* commute, in the sense that the difference of the two products is practically impossible to detect. This is why we don't see quantum mechanical effects in everyday life. The difference between writing the product one way and the other is

$$O_{mom}O_{pos} - O_{pos}O_{mom} = \hbar I,$$

where I is the identity (i.e., $\hbar I(\vec{v}) = \hbar\vec{v}$), and $\hbar$ ("h bar") is $h/2\pi$. Recall that h is (when measured in units on the human scale, $g\,cm^2/s$) the minute number 6.624×10^{-27}.

A similar situation exists for other noncommuting pairs in quantum mechanics, such as the x and y components of spin. Again, we can choose a basis in which the operator for the x component of spin is a multiplication operator, *or* we can choose a basis in which the operator for the y component of spin is a multiplication operator. (Indeed, the *spectral theorem* says that every operator in quantum mechanics has a representation in which it is simple: where it multiplies the quabability by a real function.) But we can't have both at once, and we land again on the fact that operators do not commute. That is why Heisenberg used matrices, where multiplication is not commutative.

In the case of spin, however, rather than using the Fourier transform, the appropriate tool is the Fourier series; this means that while a particle cannot simultaneously have a precise x component and y component for spin, these components can have precise values: we speak of the probability of the x or y component of spin being a particular quantity, rather than the probability of it being in a certain range. Indeed, each component of spin must always be an integer multiple of $\hbar/2$; that's why the subject is called "quantum" mechanics.

Traveling from One Function Space to Another: Wavelets and Pure Mathematics

Mathematicians say they use wavelets to have access to different *function spaces*. To get an idea of what that means, let's look at the way mathematicians have gradually broadened the definition of a function.

"Starting with the calculus and up to the 19th century, the fundamental mathematical metaphor was a function, but these functions had to have a value at every point," says Robert Strichartz of Cornell University. Before Fourier, mathematicians limited themselves to functions that can be expressed as power series, like x, x^2, x^3, ... But these analytic functions represent only a tiny fraction of all possible functions; with them, mathematicians were limiting themselves to studying a few domesticated flowers at the edge of a forest whose depths concealed unimaginably wild and strange vegetations.

Fourier's realization that one can represent even discontinuous functions as sums of sines and cosines forced mathematicians, sometimes against their will, to leave the security of the clearing to venture into the forest. There they found bizarre functions, like the one Weierstrass (1815-1897) showed "to his astounded and often indignant contemporaries" [Meyer 5, p. 12], a function everywhere continuous and nowhere differentiable, which fluctuates incessantly and whose graph is fractal, each part containing "the same complexity as the whole of the graph":

$$f(x) = \frac{1}{2}\cos 3x + \frac{1}{4}\cos 9x + \frac{1}{8}\cos 27x + \frac{1}{16}\cos 81x + \cdots .$$

223

"I turn with terror and horror from this lamentable scourge of continuous functions with no derivatives...," wrote Charles Hermite in 1893 to his friend the geometer Thomas Stieltjes [Hermite, p. 318]. But for those who mourned the honest functions of their fathers, there was worse to come. The notion of a function was to be profoundly transformed by the general definition of an integral by Lebesgue (1875-1941), the creation of function spaces (Stefan Banach, Felix Hausdorff, ...) and the theory of distributions (Laurent Schwartz, Israël Gelfand).

Lebesgue

"Lebesgue came to realize that to define a function at every point is too much of a restriction," says Strichartz. Lebesgue showed that one can talk about the integral of a "function" that does not satisfy this condition. With his more generous definition, mathematicians were forced to accept as functions such outlandish and extravagant objects as the series

$$f(x) = \cos 2\pi 10x + \frac{1}{2}\cos 2\pi 10^2 x + \frac{1}{3}\cos 2\pi 10^3 x + \frac{1}{4}\cos 2\pi 10^4 x + \cdots.$$

$$(14.1)$$

This function jumps incessantly from plus infinity to minus infinity. It gives infinity for all x that can be written with a finite number of decimals, as well as for all x that can't be written with a finite number of decimals but which, after some arbitrary point, use only some combination of the digits 8, 9, 0, and 1. It gives minus infinity for all x that after some point use only some combination of 3, 4, 5, and 6. But for almost any x one chooses, the series converges to a finite value. Although enormously many numbers give infinity—all those that a computer can manipulate, for example!—if you choose a number "at random" you will never land on those points. (In terms of probability theory, if you choose successive numbers with the help of a ten-sided die, the numbers that give infinity or minus infinity have probability zero.) "This function is totally unimaginable, yet if you want to build a reasonable mathematical theory of quantum mechanics you have to admit such functions," Strichartz says.

We see here a striking example of what can be gained by changing one's perspective: $f(x)$ is a monstrous function, giving infinity in a dense set of points; but in the space of Fourier coefficients, it becomes an innocent succession of coefficients, most of which are zero, the others forming a well-ordered progression:

$$0, 0, 0, 0, 0, 0, 0, 0, 0, 1, [89 \text{ zeros}], \frac{1}{2}, [899 \text{ zeros}], \frac{1}{3}, \ldots$$

To deal with such functions, Lebesgue defined a generalization of the integral. (For a comparison of his integral and that of Riemann, see the Appendix, p. 271.) The Lebesgue integral makes it possible to give a meaning to

$$\int_0^1 |f(x)|^2 dx,$$

which one would hesitate to describe as the area under a curve, and he showed that the Pythagorean theorem applies to it; the length squared of the function equals the sum of the squares of the orthogonal vectors that compose it. Applied to the function (14.1), this gives:

$$\int_0^1 |f(x)|^2 dx = \int_0^1 |\cos(2\pi 10x)|^2 dx + \int_0^1 \left|\frac{1}{2}\cos(2\pi 10^2 x)\right|^2 dx$$

$$+ \int_0^1 \left|\frac{1}{3}\cos(2\pi 10^3 x)\right|^2 dx + \ldots.$$

(We are using here notions discussed in *Orthogonality and Scalar Products*, p. 153). The function is a vector in infinite-dimensional space; it is decomposed in the Fourier basis, as the sum of an infinite number of cosines, multiplied by coefficients. These are also vectors. The sum of the squares of the lengths of the terms on the right equals the integral of the function squared.)

For each term on the right, the integral of the cosine squared equals 1/2, so we have the very tidy formula:

$$\int_0^1 |f(x)|^2 dx = \frac{1}{2}\left(1 + \frac{1}{2^2} + \frac{1}{3^2} + \frac{1}{4^2} + \cdots\right) = \frac{\pi^2}{12}.$$

Such a function is said to be *square integrable*—the integral of the square of its absolute value is finite.

In the 18th century, Leonhard Euler guessed, on the basis of calcu-
lations, that the sum $\left(1 + \frac{1}{2^2} + \frac{1}{3^2} + \frac{1}{4^2} + \cdots\right)$ equals $\frac{\pi^2}{6}$; he worked for
years before managing to prove it. The whole family of such sums, each
with a different exponent in the denominator, forms the *zeta function*
(ζ function):

$$\zeta(s) = \sum_{n=1}^{\infty} \frac{1}{n^s}.$$

Euler found the values $\zeta(s)$ of the ζ function for even integers s, but no
one was able to say anything about $\zeta(3)$ except that the sum is finite.
Then in 1978, mathematicians were astonished to hear that an obscure
French mathematician at the University of Caen, Roger Apéry, had proved
that $\zeta(3)$ is irrational. Initially his proof met with scepticism, even "rank
disbelief," but it was soon judged to be "quite miraculous and magnificent," and Apéry was invited to speak at the International Congress of
Mathematicians in Helsinki. "Most startling of all ... should be the fact
that Apéry's proof has no aspect that would not have been accessible to
a mathematician of 200 years ago," wrote Alfred van der Poorten [van
der Poorten].

Distributions

In the 20th century, even these bizarre functions were no longer enough,
and mathematicians added *distributions* to their arsenal. Although mathematicians had a hard time putting these objects on a rigorous basis, they
are physically quite intuitive; they make it possible to talk in a relatively
simple way about very complicated situations. Throw a rubber ball at
a wall; as the ball hits the wall it flattens slightly and then rebounds,
pushed by the restoring force of the compressed rubber. The function
expressing what happens at every point is terribly complex, and probably
irrelevant. All one really needs is the average force, as if the ball receives
an infinite impulse from the wall for an infinitely short period of time. A
distribution expresses this average.

One of the first to use distributions, under a different name, was the
Englishman Oliver Heaviside, who was dismayed by the resistance of his
contemporaries. "There was a sort of tradition," one of them said, "that

a Fellow of the Royal Society could print almost anything he liked in the Proceedings without being troubled by referees; but when Heaviside had published two papers on his symbolic methods, we felt that a line had to be drawn somewhere, so we put a stop to it" [Körner, p. 371]. But in the end mathematicians accepted these new ideas, and the theory of distributions of Schwartz and Gelfand transformed the theory of partial differential equations.

Function Spaces

A function (or a distribution) can be thought of as a point, or vector, in an infinite-dimensional function space. This perspective has encouraged mathematicians to think not just about particular functions, but about families of functions sharing certain characteristics, and thus living in the same *function space*.

A particularly important function space is the space L^2 (L in honor of Lebesgue)—the space of square integrable functions. It's a space mathematicians are fond of, since it's the space in which analogies with ordinary space work the best. Mathematicians don't have a gift of second sight that enables them to visualize infinite-dimensional space, or even space in four or five dimensions; to work in these spaces they have to learn what analogies with ordinary space are justified.

In terms of signal processing, the space L^2 is the space of functions with finite energy. It is also the space of quantum mechanics. In quantum mechanics, the state of a system is represented by a vector in L^2; energy or momentum is represented by an operator in L^2. (An *operator* treats a function the way a function treats a point: it changes a given function into another function.) Function spaces and quantum mechanics were developed at the same time, with some mathematicians—John von Neumann, Hermann Weyl—working in both fields.

Knowing to what function space a function belongs is a way of knowing its properties. More general questions can also be asked. Does an operator preserve regularity? If one gives a function to an operator, will the function that emerges still belong to the same function space? It's natural to try to attack these difficult questions by looking at a function's

Fourier coefficients, but the same properties of Fourier analysis that make it ill-suited to brief or nonstationary signals create difficulties.

One can't know anything about the local properties of a function from its Fourier transform, since a local characteristic of the function affects all the coefficients, and each coefficient contains information about the entire function. The relationship between a function's Fourier coefficients and the function space to which the function belongs is often very delicate. "If you change the coefficients in a trivial way, you can get kicked out of the function space in which you started," says Strichartz.

As in the case of our function (14.1), a very orderly Fourier series can become wildly frenzied in position space. Who, looking at our staid Dr. Jekyll, could predict that he would turn into Mr. Hyde? (Our function was carefully constructed so that we could say something about its behavior in position space. But displace the same coefficients just a little, and we would no longer know how to describe it.)

"Fourier analysis techniques were used to make enormously difficult and technical arguments," Strichartz says. "Wavelets enable us to simplify these difficult arguments, render them more comprehensible, more unified. This has applications for differential equations, mathematical physics, applied mathematics...."

As one would expect, wavelet coefficients reflect much more faithfully the local properties of a function. "Stéphane Jaffard has proved very sharp statements relating local smoothness to wavelet coefficients," Strichartz says. More importantly for pure mathematics, wavelet coefficients also help to identify the more global characteristics of a function —quantifying its regularity, for example—in order to determine what function space it lives in. One can, as Meyer wrote [Meyer 5, p. 14], travel from one function space to another, simply by changing the wavelet coefficients:

> The incessant and unimaginable fluctuations of the distributions of L. Schwartz die down as we gradually diminish the size of certain coefficients; the distribution becomes an ordinary function, more and more smooth.

We 'turn the button in the other direction' and the relief reappears, peaks rise up, yawning chasms open up... and we find again the wildest distributions. These 'voyages in space' are codified by the mathematical theory of interpolation between function spaces. A. P. Calderón, one of the pioneers of this theory, invented (using a very different language) the wavelet decomposition in order to calculate interpolations between certain function spaces.

Wavelets and Vision: Another Perspective

"Wavelet researchers tend to think that models are besides the point. We think they are the best way to approach the problem," says psychologist David Field of Cornell University. "The brain has a reason to use these codes; if we can figure it out, that may give us an idea of what wavelets will be good at."

Like Jean Morlet, researchers in vision were inspired by the work of Dennis Gabor, but coming from another direction. "Other people may have a problem they're trying to solve, like representing turbulence," says Field. "We had a solution—the visual system—but we didn't know what the problem was. What was the visual system trying to do?"

Starting in about 1965, this question was posed in terms of a debate between those who thought that the cells of the visual system react to spatial frequency and those who thought that they react instead to space—to recognize edges, for example. In 1981, the Australian mathematician S. Marcelja suggested that they do both. "He said you should really look at Gabor's paper, because Gabor describes these functions that are well localized in space and frequency," says Field. (Several years earlier Goesta Granlund in Sweden, who apparently was unaware of Gabor's work, had pointed out the significance of such functions in vision, but Granlund's 1978 paper [Granlund] "was ahead of its time and is not well known," says Edward Adelson of M.I.T.)

In 1982, Janus Kulikowski, Marcelja, and Peter Bishop suggested changing the frequency of these functions by changing the size of the window, while keeping the same number of oscillations. "Whether the window varied in size was not an issue; these things were all just called Gabor transforms, or Gaussian modulated sinusoids," says Field. Sometimes—like Morlet's *wavelets of constant shape*—they were called *self-similar Gabor functions*. But Adelson says that these techniques used in vision "were not fully developed transforms, because they were not invertible. No one knew how to actually compute an invertible transform based on self-similar Gabor functions."

Then in 1983, John Daugman and Andrew B. Watson created transforms using these functions, generalized to two dimensions, in order to model the visual system. "Not simply because it was mathematically clever," Fields says, "but because this is what the visual system does."

Seeing with "Wavelets"

In a wavelet model of vision (which Adelson says uses the "old mathematics" and thus is little indebted to the "wavelet revolution"), the image in front of our eyes is the signal. The role of the wavelets is played by the receptive fields: the patterns of light that provoke a reaction from the neurons of the primary visual cortex, at the back of the brain. These receptive fields vary in size and are oriented at different angles, different neurons responding to different receptive fields. Just as a signal in signal processing can be decomposed into wavelets of different sizes, what we look at can be decomposed into different receptive fields.

The reaction of a neuron is the "wavelet coefficient." If a neuron doesn't fire, the coefficient is zero. If it fires repeatedly and very fast, it's a big coefficient; if the response is sluggish, the coefficient is small. As in a wavelet transform, small receptive fields encode high frequencies (fine resolution, little details, good localization in space) while bigger receptive fields encode low frequencies (coarser resolution, but good localization in frequency).

"For a lot of people this seemed to be the answer: the visual system should be localized in space and frequency," Field says. "But why would the visual system want to be localized in space and frequency? Why does it use a wavelet transform?" He suggests that the object is not to compress information so that a limited number of cells respond to all visual stimuli (rather the way we see a whole range of colors with only three cones). "The principal goal of the visual system isn't to compress. Typically if I give you two patterns, even complicated patterns, you can discriminate between them; the visual system hasn't thrown out anything."

The goal, he says, is recognition. To discriminate between objects, an effective transform encodes each image using the smallest possible number of neurons, chosen from a large pool. If all visual neurons were to encode all faces, for example, recognizing the differences that distinguish one face from another would be difficult. The job is much easier if a different, small set of cells encodes the information for each face. "The goal is not to find a language with the smallest number of words, but to find a language that in any given instance allows me to describe what's out there with a small number of words."

This, Field says, is what the visual system does. At any given moment a few neurons do all the work (about one neuron in ten, according to linear models, but the visual system contains nonlinearities). But which neurons do the work changes from one moment to another; in the long run, all the neurons are about equally active. Field calls this a *sparse distributed transform* (see [Field 3]). A similar phenomenon, he suggests, may complicate attempts by Marie Farge and others to simplify the study of turbulence by determining—as Farge put it—"where we should invest lots of calculations and where we can skimp." At any one time, only a few wavelets are needed to describe the dynamically important structures in turbulence, but the wavelets that are involved in the description vary from one moment to the next. "We need methods that can take advantage of a small number of units when those units change from instant to instant," says Field.

Other researchers would call such transforms a compression system, since the encoding for a particular signal is concise. It's essentially the same idea as the idea behind the compression algorithm Best Basis (see page 259). "David Field's notion of 'sparse distribution' and our notion of 'compression' are virtually the same," says Victor Wickerhauser. "We seek to ignore all but a few big coefficients, and he seeks to ignore all but a few neurons." But Field argues that we should distinguish between "compact" transforms that encode all signals in a language with few words

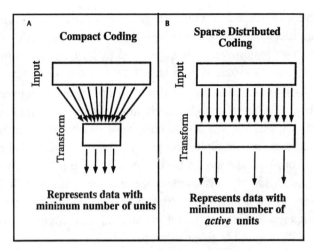

Figure 1. Compact transforms and sparse distributed transforms both compress data, but in different ways. Unlike the compact coding used when we see a range of colors using just three cones, the wavelet transform used in vision represents data with a minimum of active wavelets; which wavelets are used changes from instant to instant. This is also the case when wavelets are used to represent dynamically important structures in turbulence, says David Field. He believes new methods are needed to take advantage of such transforms. (Courtesy of David Field.)

(sometimes called "dimensionality reduction"), and "sparse distributed" transforms that encode each message in few words, chosen from a large vocabulary (see Figure 1).

Why Wavelets?

Even if a sparse distributed transform is desirable for recognizing what one sees, why should the visual system use wavelets? Field says the answer lies in the signal. Too often, he says, theories about the visual system have been based on the response of visual neurons to straight lines, checkerboard patterns, and random points. It is more reasonable, he thinks, to study the scenes found in nature. "If I give you ten white noise patterns and then give you a new one and ask, Have you seen this one?, you'll be awful. But I can give you ten thousand natural scenes and then say, Have you seen this one?; you'll say, 'Oh yeah, I remember that one.' We're extremely good at coding that kind of information."

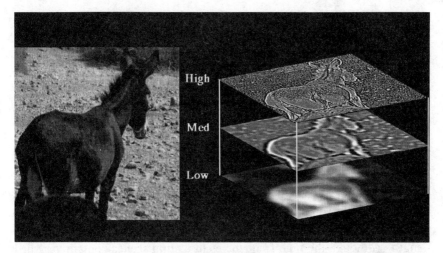

Figure 2. A "natural scene" and its decomposition into high, middle, and low frequencies, using a two-dimensional wavelet transform. The effectiveness of a transform cannot be understood or evaluated in isolation; it depends on the relationship between the transform and the properties of the data to be encoded. Natural scenes have certain statistical properties in common. For example, as shown in the above decomposition, they are redundant: many edges found in low frequencies also exist in middle or high frequencies. Wavelets with a certain narrow bandwidth of frequencies and certain narrow range of orientations can encode natural scenes concisely, with just a few coefficients. (Courtesy of David Field.)

Natural scenes are very diverse, but they represent a minute proportion of all possible scenes, and they have certain statistical properties in common, Field says (see [Field 1] and [Field 2]). On average, for example, if one makes a Fourier transform of a natural scene, the amplitude will decrease roughly as the inverse of the spatial frequency (and the energy will decrease as $1/f^2$: each "octave" has the same energy). This means that contrast—variation in the intensity of pixels—doesn't change with scale; it doesn't matter whether you are looking at something close up or from a distance. In addition, natural scenes are redundant. Many edges found in low frequencies also exist in middle or high frequencies, as seen in Figure 2; this has to do with the alignment of phases.

A wavelet transform of a scene with these properties has a few big coefficients and many that are zero. "That is what wavelets do to natural phenomena," Field says: they arrange a signal neatly in a few compartments, while an encoding of the same scene according to pixel intensity would be much more diffuse. But not just any wavelet will do.

Which Wavelets?

The typical "wavelet transform" used in vision is not orthogonal, says Field; while the invertibility of orthogonal transforms is useful in many signal processing applications, the visual system doesn't need to invert. Nor does he favor normalized wavelets, which all have the same vector length. Although the high frequencies in natural scenes contain very little "energy," we are extremely sensitive to high frequencies, Field says. "As the energy falls off, you want the vectors to increase in their sensitivity; you want the vector length to increase in proportion to frequency. This is something we're working on now."

The "wavelets" of the visual system are also characterized by the range of frequencies—the bandwidth—to which they react, typically about 1.5 octaves. The visual neurons are also narrowly tuned to orientation. These are the bandwidths of frequency and orientation needed to encode a natural scene with just a few neurons, Field says. "It seems that our visual system has evolved to get this particular kind of information out of our world."

With these bandwidths, the receptive fields seem to exploit the redundancy found in natural scenes. "Structure in real scenes is only predictable for a small range of frequencies in a small range of orientations," says Field. "That's the reason you have limited bandwidth in orientation and spatial frequency—the information in natural scenes is predictable within that bandwidth. If the world were made up of long straight edges, information would be completely predictable across scale, and you would want to use very broadband functions. If the information across scale was extremely unpredictable, then you would want functions with very narrow bandwidths—something like a Fourier transform. But the world is predictable in the range of one to two octaves."

Steerable Filters and Shiftable Transforms

This picture of wavelets taking advantage of the statistical properties of natural scenes in order to promote recognition is interesting, "but I don't think it's the whole story by any means," remarks Adelson. "We also have to ask how these cells' outputs are used to do things like analyzing texture, motion, orientation, contours.... Anyone who has tried building a computer vision system realizes that the representations in the early stages are very important in determining what is possible in later stages."

"One interesting point is that mammalian vision systems always use filters that are oriented in two dimensions. There is little use for the criss-cross diagonal filters that you get by separately combining one-dimensional filters—although these are the ones commonly used when wavelets are generalized to two dimensions." He and William T. Freeman have shown how to design *steerable filters* that allow one to determine the response of a filter of any orientation without explicitly applying that filter; instead, one uses a few filters corresponding to a few angles and interpolates (see *Wavelets in Two Dimensions*, p. 193, and [Freeman, Adelson]).

Another important property for vision applications is *shiftability*, Adelson says—an extension of the concept of translation invariance. A *jointly shiftable* transform may be invariant with respect not only to position of the input signal but also to scale; or it may be invariant with respect to position and orientation (see [Simoncelli et al.]). "These properties of steerability and shiftability cannot be achieved with critically sampled transforms," Adelson says, "so we advocate transforms that are heavily oversampled."

Which Wavelet?

The would-be user of wavelets faces an embarrassment of riches. Even excluding hybrids like wavelet packets, choosing a wavelet may seem overwhelming. Not only are there two big classes of wavelet transforms—continuous and discrete—but discrete transforms can be redundant, orthogonal, or biorthogonal. Each category contains innumerable possibilities, Daubechies' wavelets alone constituting a very big class.

One extreme is to reach each time for the familiar. "Everyone uses the same thing: the Mexican hat, the Morlet wavelet, and Daubechies' wavelets," complains Marie Farge, researcher at the Ecole Normale Supérieure in Paris.

Another is to get so caught up in the search for new bases that the original purpose is forgotten. "The discovery of filter banks and wavelet bases has created a popular new sport of basis hunting. Families of orthogonal bases are created every day. This game may however become tedious if not motivated by applications," writes Mallat in the preface to his book, *A Wavelet Tour of Signal Processing* [Mallat 4, p. xvi].

How should one choose the right wavelet for a particular application? Farge cautions against spending "infinite energy choosing a wavelet. You shouldn't get caught in the paradox that for a given signal there is one optimal wavelet. It's true, that is the case, but then you miss the whole point of a general method like wavelets. If you really want to find the best way to analyze a signal for a given application, then you're better off doing something else."

Olivier Rioul of the Ecole Nationale Supérieure des Télécommunications in Paris, who wrote his doctoral thesis on wavelets and image

239

compression, disagrees. "In general people say that any halfway decent wavelet will do. But there are differences," he says. Why not try to find the wavelet best adapted to a given task, he asks, just as Best Basis seeks the basis that is best adapted to a signal? He suggests custom-making one's own wavelets, with the properties one wants.

Unfortunately, rephrasing the question in terms of "what properties" rather than "what wavelet" doesn't completely solve the problem. There are elements of an answer; for example, in image compression, regularity is important and frequency selectivity is not (see Figure 1), while in audio compression, frequency selectivity is crucial. "But we still don't really know what properties play what roles," Rioul says.

In general there are two kinds of choices to make: the system of representation (continuous, discrete, ...) and the properties of the wavelets themselves: for example, the number of vanishing moments and the degree of regularity. The common theme in all these choices is *trade-off*— hardly surprising in a field built on the limits imposed by the uncertainty principle. If you want more resolution in frequency, you get less in time; if you want more vanishing moments you must increase the size of your wavelet's support, which makes for more computations ....

The System of Representation

Orthogonality confers conciseness and speed at the price of arbitrary scales and some difficulty in analysis, especially for recognizing patterns, since orthogonal representations are not translation invariant. Continuous transforms have, in general, the inverse advantages and disadvantages.

If speed is the priority, then an orthogonal transform is generally best. "A decomposition ... using Mallat's fast algorithm is typically orders of magnitude faster than a redundant analysis, even if one uses the fastest available algorithms," write Michael Unser of the Swiss Federal Institute of Technology in Lausanne and Akram Aldroubi of the National Institutes of Health [Unser, Aldroubi, p. 627].

But it may be worth sacrificing some speed to gain some other feature. "I am only aware of one application for which orthogonality is desirable,

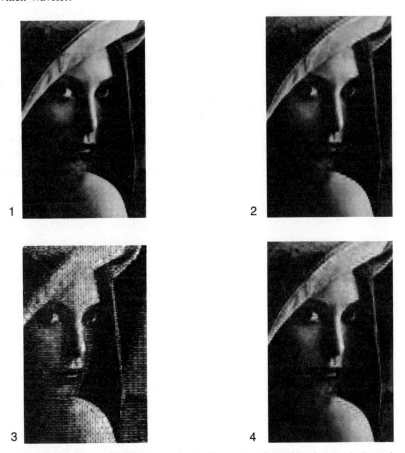

Figure 1. The quality of a reconstructed image is affected by the regularity and frequency selectivity of the wavelets used to compress it. Three different filters of length 8 (corresponding to three different wavelets) were used to compress the same image. (1) The original image. (2) Image reconstructed after compression with the Daubechies wavelet that, among Daubechies wavelets, is the most regular and the least selective in frequency for the length of the support. (3) Image reconstructed after compression using a wavelet very selective in frequency, with negative regularity. (4) Image reconstructed after compression using a wavelet intermediate between the other two. The images are all zooms on a portion of a digital image; without the zooms, images (1), (2), and (4) would not look very different. The small visible squares correspond to pixels of the digital image and should not be interpreted as a defect of the processing. (Courtesy of Olivier Rioul and the Centre National d'Etudes des Télécommunications (CNET) in France.)

namely, image coding for data compression. In all other image processing applications, overcompleteness is advantageous," says Edward Adelson of MIT.

It is possible to achieve translation invariance with a "semi-continuous' transform, in which the scale parameter (but not the translation parameter) is discretized and sampled. A fast algorithm exists for these highly redundant transforms: the *algorithme à trous* (*algorithm with holes*), developed by M. Holschneider, Richard Kronland-Martinet, Jean Morlet, and Philippe Tchamitchian [Holschneider et al.]. Another scheme achieves translation invariance while limiting redundancy by translating the sampling grid when the signal is translated [Mallat 4, p. 151].

Discrete Transforms and Frame Theory

Frame theory provides a language for describing the completeness, stability, and redundancy of a whole variety of discrete transforms—including orthogonal and biorthogonal transforms as special cases. It was originally developed in 1952 by Duffin and Schaeffer [Duffin, Schaeffer]; more recently other researchers, notably Ingrid Daubechies, have shown how it applies to windowed Fourier analysis and wavelets.

A frame is a family of vectors such that the scalar products of those vectors with a signal can be used to represent the signal. Such a frame is more or less *tight* depending on how well the "energy" of the signal is preserved by the decomposition; if it is preserved exactly and the frame vectors have a unit norm, then the frame is an orthonormal basis.

Biorthogonal Wavelets

Biorthogonal wavelets provide perfect reconstruction of the signal, without redundancy, using two sets of wavelets: one for the decomposition, another for the reconstruction. Generally, there is more freedom when designing biorthogonal wavelets—particularly important in higher dimensions. For example, one can use the "lifting scheme" to construct new biorthogonal wavelets, but not new orthogonal wavelets.[1]

---

[1] As discussed in *The Lifting Scheme*, p. 257, given a known orthogonal wavelet, it is possible to figure out, after the facts, how to create it using lifting.

In addition, biorthogonal wavelets can be both symmetric and have compact support. (Symmetry is incompatible with orthogonality except in the case of the Haar wavelet, or by using more than one scaling function; see *Multiwavelets*, p. 201.) For some applications, such as numerical analysis, symmetry is not important. In image processing, symmetric or antisymmetric wavelets make it possible to use a "folding technique" to avoid artificially large coefficients at the boundaries. It is also thought that quantization errors are less evident when symmetrical wavelets are used. This effect is not well understood; in his research Rioul did not find that symmetry had a significant effect on the usual criterion for distortion used in judging the quality of reconstructed images (the PSNR or *peak signal over noise ratio*) [Rioul, p. 109]. "So if symmetry does have a role to play, it's in connection with the subjective perception of visual quality, which is difficult to evaluate," he says.

As usual, these benefits come at a price. Biorthogonal wavelets bases do not reproduce the signal energy exactly and reconstructing a signal from these coefficients may amplify any error introduced on the coefficients. This is an issue in compact signal coding where the quantization error should not be amplified by the transform. In this case, the "nearly orthogonal" biorthogonal wavelets designed by Albert Cohen, Ingrid Daubechies, and J.-C. Feauveau are often used.

Choosing a Basis Adaptively

Why not automate the choice of basis, giving a signal to an algorithm like Best Basis or Matching Pursuit, and letting it decide what wavelets, wavelet packets, or other waveforms to use, based on the characteristics of the signal? Often this is a good approach, but there are potential drawbacks.

One is that if the signal noise dominates the signal, there is a risk of choosing a basis that fits the noise, which degrades the signal estimation. Another is that if an adaptive scheme like Best Basis is used for compression, the coding for the compressed signal has to indicate what basis was used. For example, for a signal of size n, the Best Basis dictionaries contain more than $2^{n/2}$ bases; if all bases are equally likely to be chosen, it takes more than $n/2$ bits to encode the choice of basis. Often this overhead wipes out the gain obtained by optimizing the basis choice.

Properties of Wavelets

For the many applications of wavelets that exploit their ability to encode a signal with just a few non-zero coefficients, the ideal wavelet is one that will encode a signal using the greatest possible number of zero coefficients, or coefficients close to zero. This happens if most of the fine scale wavelet coefficients are small; in turn this depends largely on the regularity of the signal, the number of vanishing moments[2] of the wavelet, and the size of its support.

Vanishing Moments and Length of Support

The number of vanishing moments, which is weakly linked to the number of oscillations (the more vanishing moments a wavelet has, the more it oscillates), determines what the wavelet doesn't "see." A wavelet with one vanishing moment doesn't see linear functions: the wavelet coefficient—the scalar product of the wavelet and the function—is zero. (The two are orthogonal to each other.) Two vanishing moments make a wavelet "blind" to quadratic functions as well; three vanishing moments, to cubic functions; and so on. A wavelet with many vanishing moments also gives small coefficients when it is used to analyze low frequencies; the coefficients measure the resemblance between a wavelet and the signal, and a wavelet that oscillates rapidly doesn't look like slow low-frequency waves.

The practical effect of vanishing moments is to concentrate the information of the signal in a relatively small number of coefficients. This

---

[2]The *kth moment* of a function f is the integral of the function multiplied by its variable raised to the power k:

$$m_k = \int_{-\infty}^{\infty} f(x)x^k \, dx.$$

The moment "vanishes" if this integral is zero.

can be useful for compression, as well as for analyzing signals with singularities and discontinuities: the unpredictable variations of the signal give big coefficients that stand out against a background of zero or small coefficients. But as Ingrid Daubechies proved, if a wavelet has p vanishing moments, then its support must be at least $2p - 1$, and increasing the size of the support increases the number of computations. How many vanishing moments are desirable depends on the application. "For the numerical analysis applications of Beylkin, he uses five or six vanishing moments, and he really does need them," says Daubechies. "Of course, in practice, the moment doesn't really need to vanish completely; it is often sufficient if it is very, very small."

For image compression, wavelets with three or four vanishing moments are most often used. In work with magnetic resonance imagery and positron emission tomography, for example, Unser generally uses cubic spline wavelets with four vanishing moments, "because they provide a good compromise in terms of the signal-to-noise ratio and smoothness of the reconstruction versus ringing."

In deciding whether increasing the number of vanishing moments is worth the cost, the signal to be encoded must be considered. If it is very regular except for a few isolated singularities, then it may be best to choose a wavelet with many vanishing moments, at the cost of increasing the size of the support [Mallat 4, p. 243]. But for a signal with more singularities, a wavelet with a smaller support may be a better choice.

Among orthogonal wavelets, Daubechies' wavelets provide the smallest possible support for a given number of vanishing moments. A better trade-off between support size and vanishing moments can be achieved with multiwavelets, but decomposing a signal with multiwavelets is slightly more complicated than a standard orthogonal wavelet decomposition.

Regularity

Not everyone is enamored of vanishing moments. "From the point of view of encoding, of signal processing, I don't really see what's the point of

vanishing moments, except as a necessary condition for regularity," says
Rioul. The *order of regularity* of a wavelet is the number of its continuous
derivatives; to have a regularity greater than n, a wavelet must have at
least $n + 1$ vanishing moments.

Rioul considers regularity the most important contribution of wavelets
in the field of encoding. Encoding a smooth picture with a discontinuous
function like the Haar wavelet produces a discontinuous picture; edges
appear that don't exist in the original picture. Using a regular (smooth)
function as the analyzing function can prevent such artifacts.

In image processing Rioul has found that increasing regularity, up to a
certain point, improves the results, independent of the number of vanishing
moments. His research suggests that for image encoding, having wavelets
that are one or two times continuously differentiable is sufficient. For
a given number of vanishing moments, the most regular wavelets are
splines. (Splines are piecewise polynomial functions; they have long been
used in such fields as interpolation and curve fitting, computer graphics,
and computer aided design.)

Selectivity in Frequency

In Fourier analysis, the analyzing function is a sinusoid of precise frequency, which gives a coefficient that refers to that frequency and no
others. But a wavelet is composed of a mixture of frequencies, shown by
its own Fourier transform; wavelet coefficients necessarily refer to this
mixture of frequencies. The narrower this range of frequencies, the more
selective the wavelet is in frequency: it cuts the signal into frequencies
more precisely. (In practice, with the fast wavelet transform, wavelet coefficients are computed using a filter associated with the wavelet, so one
is really concerned with the frequency selectivity of the filter.)

Historically, "frequency selectivity" and "filter" were essentially synonymous; a filter was defined as a device that allows some frequencies
to pass while blocking others. The Fourier transform was thus the ultimate filter. The central problem of signal processing was how to analyze
a signal simultaneously in frequency *and* time, to accommodate the obvious fact that if we were to listen to the Fourier transform of a piano

sonata—all notes (frequencies) played simultaneously, some louder than others—we would feel that the music had lost a lot in translation.

But researchers experimenting with wavelets soon found that for image processing, posing the problem in terms of space/frequency localization was not the most effective approach.

"For image processing you really do not care about having a wavelet that is well localized in frequency because the image structures (edges, for example) are very poorly localized in frequency," says Mallat. "There the ability to compress depends rather on vanishing moments, and visual quality depends on the wavelet regularity. On the other hand, for speech, frequency localization is very important because speech structures are well localized in frequency. In the past, most 'signal processors' were 'speech processors.' This is why they initially tended to see performance as depending essentially on the frequency/space localization of images. This is also why nobody tried to do what Daubechies did, namely optimize the number of vanishing moments and not the frequency localization of the filters."

In audio compression, good frequency localization is crucial, as it masks small errors due to quantization. This good frequency localization is obtained by using many vanishing moments,[3] at the cost of a large support. D. Sinha and A. Tewfik [Sinha, Tewfik] found that increasing the number of vanishing moments up to 30 improved the perceived quality. But the resulting filters have at least 60 nonzero coefficients, compared to only two for the filter associated to the Haar wavelet.

When optimal time-frequency localization is the goal, the irreducible obstacle is of course the Heisenberg uncertainty principle, setting a lower limit to the product of the uncertainty in time and the uncertainty in frequency. The functions that approach this limit most closely are the

---

[3]"The vanishing moments impose a condition on the Fourier transform in the neighborhood of 0: that $\hat{\psi}(\omega) = O(\omega^p)$ for ω small if it has p vanishing moments," says Mallat. (In that formula, O means "at most on the order of.") "The frequency localization says that the wavelet should be negligible for $|\omega|$ outside an interval $[\omega_0, \omega_1]$. Clearly, the vanishing moments influence the fact that it is small for $|\omega| < \omega_0$ but it does also influence (for orthogonal wavelets, and this is less evident) the fact that $\hat{\psi}(\omega)$ is small for $|\omega| > \omega_1$. In fact this is related to the fact that more vanishing moments gives more regularity."

Gabor functions, but the Gabor transform is redundant and its inversion is unstable. Wavelets that approximate Gabor functions, and thus are also time-frequency optimal, while providing concision and stability, have been found by Michael Unser and his colleagues [Unser et al. 2]. These are the biorthogonal B-spline wavelets, whose corresponding scaling functions are the B-splines, a special group of splines. The B-spline wavelets converge to Gabor functions as the degree of the splines increases (1 for piecewise linear splines, 2 for arcs of parabolas ...). "The joint localization (or uncertainty product) of the cubic B-spline wavelets is already within 2 percent of the Heisenberg limit," Unser says, which "should be sufficient for most practical applications." However, these wavelets do not have compact support, so one cannot do exact calculations with them.

Different Transforms: A Summary

> A very common pitfall when using any kind of transform is to forget the presence of the analyzing function in the transformed field, which may lead to severe misinterpretations, the structure of the analyzing function being interpreted as characteristic of the phenomena under study.

—Marie Farge, *Wavelet Transforms and Their Applications to Turbulence*, p. 429

The Fourier Transform

Kind of Decomposition	Frequency
Analyzing Function	Sines and cosines, which oscillate indefinitely
Variable	Frequency
Information	The frequencies that make up the signal
Suited for	Stationary signals (predictable: obeying constant laws)
Notes	With the fast Fourier transform (FFT) it takes $n \log n$ computations to compute the Fourier transform of a signal with n points.

The Windowed Fourier Transform

Kind of Decomposition	Time-frequency
Analyzing Function	A wave limited in time, multiplied by trigonometric oscillations. The size of the wave, or "window," is fixed for each analysis, but the frequency inside the window varies.
Variables	Frequency; position of the window
Information	The smaller the window, the better time information one has, at the cost of losing information about low frequencies; large windows give better frequency information but less precision about time.
Suited for	Quasi-stationary signals (stationary at the scale of the window)
Notes	Sometimes this is called "short-time Fourier analysis," or, when a Gaussian is used as the envelope of the window, the "Gabor transform." While the Fourier transform is orthogonal, the most obvious forms of windowed Fourier cannot be orthogonal.

The Wavelet Transform

Kind of Decomposition	Time-scale
Analyzing Function	A wave limited in time, with a fixed number of oscillations; the wavelet is contracted or dilated to change the size of the "window" and thus the scale at which one looks at the signal. Since the number of oscillations does not change, the "frequency" of the wavelet changes as the scale changes.
Variables	Scale; position of the wavelet
Information	Small wavelets provide good time information but poorer frequency information. Large wavelets provide good frequency information but poorer information about time.
Suited for	Nonstationary signals, such as very brief signals and signals with interesting components at different scales (fractals, for example).
Notes	The wavelet transform can be continuous or discrete; orthogonal and biorthogonal wavelets are special cases of the discrete wavelet transform. With orthogonal wavelets, the transform of a signal with n points can be computed with cn computations; the value of c depends upon the complexity of the wavelet.

Malvar Wavelets (Adaptive Windowed Fourier)

Kind of Decomposition	Time-frequency-scale
Analyzing Function	A curve, limited in time, of special shape, multiplied by certain trigonometric oscillations
Variables	Frequency, position of the "window," and size of the window, all of which can change independently
Suited for	Signals for which one is primarily interested in the dynamics of the signal as a function of time (music, speech), since one can cut the signal in a nonuniform way (adaptive time segmentation)
Notes	With three independent parameters (position, frequency, and scale), Malvar wavelets form a redundant system. But one can easily extract from it an infinite number of orthogonal bases, which can be used in the compression algorithm, Best Basis, which seeks the best basis suited to a given signal.

Wavelet Packets

Kind of Decomposition	Time-frequency (with the possibility of also analyzing by scale)
Analyzing Function	Looks like a wavelet multiplied by trigonometric functions
Variables	Frequency, position, and the possibility of varying scale as well.
Suited for	Signals that combine nonstationary and stationary characteristics, such as fingerprints.
Notes	With three independent parameters (position, frequency, and scale), wavelet packets form a redundant system. But one can easily extract from it an infinite number of orthogonal bases, which can be used in the compression algorithm, Best Basis, which seeks the best basis suited to a given signal. Wavelet packets are far more flexible than wavelets, but less is known about the interpretation of the coefficients.

Matching Pursuit

Kind of Decomposition	Time-frequency-scale
Analyzing Function	A Gaussian of variable window size, multiplied by trigonometric functions
Variables	Frequency, position of the "window," and size of the window, can change independently
Suited for	Highly nonstationary signals composed of very different elements
Notes	Encodings of highly nonstationary signals by Matching Pursuit are concise and translation-invariant. Encoding a signal with n points requires more computations than Best Basis, but reconstruction of the signal is fast.

Wavelets, Music, and Speech

Windowed Fourier analysis was developed by Dennis Gabor in the context of acoustics. "This type of analysis is in constant use in speech and acoustics; it's a fundamental tool," says Christophe d'Alessandro of the French Centre National de Recherche Scientifique, who works at the University of Paris-Sud at Orsay.

Has multiresolution wavelet analysis been an important development in this field? The first attempts to analyze speech with orthogonal wavelets were "a total failure. Researchers couldn't interpret the wavelet coefficients," says Yves Meyer. Today researchers in speech and acoustics give a somewhat more positive assessment. "In speech recognition, for the moment wavelets haven't produced anything noteworthy, but there have been interesting results in pitch recognition in speech," says d'Alessandro.

He suggests that the apparent lack of new applications of wavelets in speech and acoustics is at least partially due to the fact that "the same kind of techniques, without the name 'wavelet,' already long existed in these fields" (for example, in the form of the quadrature mirror filters used in speech, in addition to older techniques). "In the short term, wavelets gave a new impetus to these techniques, but in fact they were already well known and exploited; transforms like wavelet transforms have been fundamental for some 50 years. It's thanks to them that modern theories about the production, perception, and transmission of speech were born."

Victor Wickerhauser points out, however, that the original quadrature mirror filters, which weren't very regular, produced annoying artifacts when they were used with the number of iterations necessary to obtain high compression ratios. Daubechies's regular wavelets made more ef-

253

ficient decompositions possible. (But the wavelets used in speech are not necessarily the same as those used in other applications. "They are wavelets that are very well localized in frequency, very different from the wavelets in image processing," says Mallat; see *Which Wavelet* p. 247.)

For speech, frequency localization is very important because the speech structures are well localized in frequency, and because of the "masking" properties of hearing. If a strong signal has a narrow frequency support, it will mask a small amplitude signal having a close frequency, which means that we will not hear this weaker signal. When audio signals are recorded, the small errors introduced by quantization may not be heard because of this masking effect (see for example [Mallat 4, p. 505]).

Wavelets are also used in the study of music. Richard Kronland-Martinet and others in Marseille use wavelets to analyze and synthesize musical sounds; Neil Todd (of City University, London University) uses wavelets to analyze rhythm.

Mathematician Ronald Coifman of Yale University first began applying wavelets to the study of music when he collaborated with Jonathan Berger, then director of Yale's Center for Studies in Music Technology, on a restoration of a historic recording of Brahms (see Chapter 5). Since then they have applied techniques similar to those used in the Brahms denoising to pitch tracking and timbre separation [Berger et al.]. More generally they have been trying to bridge what Berger calls "the traditional schism between music analysis at the signal level and the pattern and structural analysis of score representations of music."

Very early on we came to realize that 'finding the music in the noise' cannot be achieved by raw computation and fancy signal processing only, but must include considerations of higher level musical structure. Specifically, we consider reductive analytical techniques and compare these methods to signal processing methods that facilitate selective recomposition of the signal. Our principal construction encompasses a wide range of apparently incompatible representations, including traditionally notated scores, but also raw digital audio of a specific performance. In the low-level domain of digital audio analysis we confront the problem of polyphonic pitch-tracking and timbral separation.

In the high-level domain of score and symbol we deal with issues of pattern and structural hierarchy. Our approach aims at filling the apparent void between these two remote strata with intermediate levels of partial reconstruction of one or the other from their respective elements.

The Lifting Scheme

The "lifting scheme" introduced by Wim Sweldens [Sweldens] produces wavelets that a few years ago would not have satisfied any definition of wavelet. Although "wavelets" was already quite an elastic term, encompassing the wavelets of both orthogonal and continuous transforms, such wavelets share at least something in common: they are formed by translations and dilations of a mother wavelet. These "first generation" wavelets can be used to analyze data that naturally occur on a line (music or an electrocardiogram, for example) or on a plane (a photograph). The Fourier transform can be used to construct such wavelets, since in Fourier space translation and dilation become algebraic operations.

The motivation for lifting, says Sweldens, was to find a way to make wavelets that can be used in a wider variety of situations: for example, "irregular samples and triangulations, complex regions, surfaces, manifolds, adaptive and non-linear transforms."

In these cases, Sweldens says, translations and dilations can't be used, so the Fourier transform is of no help. The basic idea behind lifting is to take a simple or even trivial wavelet transform and gradually improve ("lift") properties such as smoothness and vanishing moments. The important realization was that translation and dilation are not essential to building wavelets with the desired properties. "The notion that a basis function can be written as a finite linear combination of basis functions at a finer, more subdivided level, is maintained and forms the key behind the fast transform," he writes in a paper with Peter Schröder [Schröder, Sweldens]. "The main difference with the classical wavelets is that the filter coefficients of second generation wavelets are not the same through-

out, but can change locally to reflect the changing ... nature of the surface
and its measure."

Schröder and Sweldens have shown how to use the lifting scheme to
produce spherical wavelets, with potential in geophysics and geography,
computer graphics (virtual reality), and astronomy. The Jet Propulsion
Laboratory is considering using spherical wavelets to analyze solar winds.

In such settings lifting will only generate biorthogonal transforms,
which use two sets of wavelets, one for decomposition and one for reconstruction. But using some 2,000-year-old algebra (the Euclidean algorithm), all existing "first generation" wavelets can be built using lifting
(for an overview, see for example [Daubechies, Sweldens]). So far this
process can only be done backwards. "Given an orthogonal wavelet, we
can after the fact figure out which lifting steps would have led to the
same wavelet," Sweldens says. "However, we don't know at this point
how to build an orthogonal wavelet from scratch with lifting."

"Originally I never thought lifting would be useful in this context.
To my surprise this turned out to be not true," he adds. "Lifting has
many advantages in the first generation setting: in-place computation,
integer-to-integer transforms (important for lossless coding), zero-delay
filter banks, and a speed-up of a factor two over the standard fast wavelet
transform."

The improvement in speed comes from making optimal use of the
close relation between the high-pass and low-pass filters in the wavelet
transform. This allows both filters to be applied roughly at the cost of
one.

Best Basis

Choosing a basis in which to decompose a signal means selecting a certain compromise between time and frequency. This choice can be represented in terms of *Heisenberg boxes*. Each box corresponds to one element of the basis and roughly represents the time-frequency information it encodes. A Heisenberg box (also called a "cell") is a rectangle in the plane whose sides are respectively the range of values and the range of frequencies represented by the basis element. The Heisenberg uncertainty principle sets a lower bound to the size of these boxes: the product $\Delta t \Delta \tau$ must be at least $1/2\pi$. One can think of the elements of an orthogonal basis— for example, all the translates and dilates of a given mother wavelet— as *paving* the time-frequency plane with Heisenberg boxes: the boxes covering the plane without overlapping.

This image is somewhat misleading. It is not true that the Heisenberg boxes corresponding to an orthogonal basis cover the entire time-frequency plane without overlapping. Nor, indeed, is the picture of time-frequency boxes of different shape but the same area really correct; most time-frequency functions don't reach the lower bound set by the Heisenberg uncertainty principle, and some get closer than others. Further, there is no reason to expect the time-frequency information corresponding to a function to take the form of a neat rectangle; an irregular blob probably represents it better.

But for computations, this representation is useful. "The plane of Heisenberg boxes is an easily computable idealization of the time-frequency plane; the latter is rigorously defined but difficult to work with," says Victor Wickerhauser of Washington University in St. Louis.

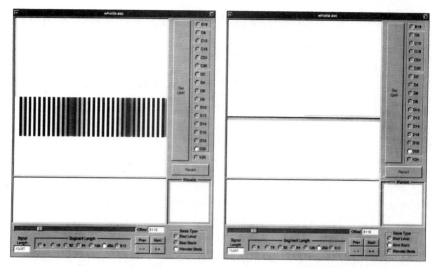

Figure 1. A recording (at 8012 samples per second) of a person whistling. Left: the signal represented in a wavelet basis. Right: the same signal in its "best basis," using wavelet packets. The wavelet basis at left localizes the frequency only within an octave, while the best basis analysis shows that the frequency falls in a much narrower band. (The vertical stripes at left could be used to better localize the frequency, but the best basis decomposition does this automatically.) (Courtesy of M. Victor Wickerhauser.)

Given a signal, the compression algorithm Best Basis chooses, from a library of bases, the basis in which the signal can be represented with the least possible area in this plane; two examples are shown in Figures 1 and 2. This is done by choosing the basis that minimizes a *concave cost function.* In the original conception of Best Basis, the cost function used was the entropy function. (But while entropy in communication theory gives a lower bound to the number of bits needed to encode a signal, the Best Basis entropy measures the energy distribution of a signal.) In applications like noise removal, other cost functions are often used; the only requirement is that they be concave.

Computations are fast and automatic; a computer program reads signals and displays them as Heisenberg boxes, which can be colored to indicate the size of coefficients. Only significant coefficients are considered.

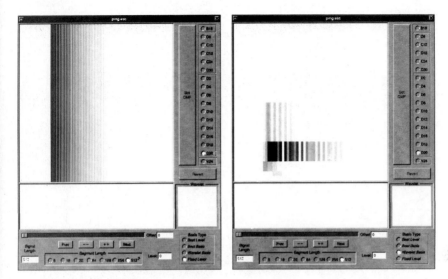

Figure 2. A damped oscillator that receives an impulse, shown at left in the Dirac basis, at right in a wavelet basis. The signal can be represented yet more concisely using wavelet packets; that decomposition is not shown here. (Courtesy of M. Victor Wickerhauser.)

"The resulting pictures are useful in measuring certain signal characteristics and in recognizing certain types of transient signals like mixtures of chirps," Wickerhauser says. "Also, the correspondence between geometry (disjoint covers) and algebra (orthonormality) has resulted in certain insights about decompositions of functions and in combinatorial methods for computing the complexity of our fast algorithms." (For more detail, see [Coifman, Wickerhauser 1], [Coifman, Wickerhauser 2], or [Wickerhauser 2]. For potential drawbacks of all algorithms that choose a decomposition adaptively, see *Which Wavelet*, p. 243.)

PART III
Appendices

Mathematical Symbols

Δ, δ	delta
ϵ	epsilon (commonly used to represent small numbers)
θ	theta (a variable angle)
μ	mu (often represents measures)
ξ	xi (a variable representing the wave number, or spatial frequency, for Fourier transforms of signals that depend on x, representing space)
σ	sigma (standard deviation)
τ	tau (a variable representing frequency, for Fourier transforms of signals that depend on t, representing time.)

Note: k often replaces ξ or τ in formulas for Fourier series, according to the mathematical convention that k represents integer variables, while ξ and τ represent continuous variables.

ϕ or φ	phi (the scaling function; also used to represent the phase angle)
ψ	psi (the wavelet; in quantum mechanics, the wave function)
$\sum$	sum
$\int$	integral
$\prod$	product
$\cup$	union
$\cap$	intersection
∞	infinity

A Review of Some Elementary Trigonometry

The trigonometric functions we use in this book are sine (abbreviated sin) and cosine (abbreviated cos). Consider a unit circle (a circle with radius 1) centered at 0: $x^2 + y^2 = 1$. Starting from the point (1,0), go a distance θ counterclockwise, as shown in Figure 1. The coordinates of the point where you stop are $(x, y) = (\cos\theta, \sin\theta)$.

We have just defined the sine and cosine of an arc of circle. We can also consider sines and cosines as functions of the angle corresponding to the arc. The angle is then usually measured not in degrees but in radians, as in the above

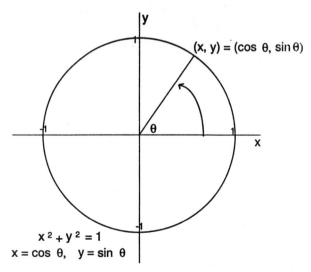

Figure 1. The sine and cosine of an arc of circle of radius 1, centered at 0.

267

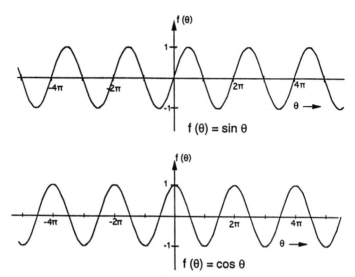

Figure 2. Above: the graph of $\sin\theta$. Below: the graph of $\cos\theta$. Both functions are periodic of period 2π. The two graphs are identical, shifted in phase.

construction: the angle θ is measured by the length of the corresponding arc of unit circle. A full turn around the circle equals $360° = 2\pi$ radians. A $90°$ angle thus equals $\frac{\pi}{2}$ radians. Often the word "radians" is omitted.

For example, if $\theta = 90° = \frac{\pi}{2}$, then $\sin\theta = 1$ and $\cos\theta = 0$. If $\theta = 360° = 2\pi$, then $\sin\theta = 0$ and $\cos\theta = 1$.

It follows from the definition that

$$(\sin\theta)^2 + (\cos\theta)^2 = 1.$$

Often, $(\sin\theta)^2$ is written $\sin^2\theta$:

$$\sin^2\theta + \cos^2\theta = 1.$$

Graphing Sines and Cosines

We can graph $\sin\theta$ and $\cos\theta$ by "unrolling" θ from the circle and representing it on the horizontal axis (2π equals one complete revolution, 4π equals two revolutions, etc.). Figure 2 shows $\sin\theta$ and $\cos\theta$, each periodic of period 2π (i.e., the functions repeat themselves every 2π).

The same graphs could be relabeled to show $\sin 2\pi\theta$ and $\cos 2\pi\theta$, each periodic of period 1. We could also change our variable: θ could just as well be x or t.

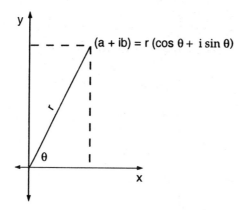

Figure 3. A complex number can be written in Cartesian notation as $a + ib$, or in trigonometric notation as $r(\cos\theta + i\sin\theta)$, where $i = \sqrt{-1}$. The numbers r and θ are then the polar coordinates of the point (a, b). Adding complex numbers is easier with Cartesian coordinates, and multiplication is easier with polar coordinates.

The Complex Plane

Now let's consider sines and cosines in the complex plane, shown in Figure 3. Then the complex number $a + ib$ equals $r(\cos\theta + i\sin\theta)$, and we see that Euler's equation $e^{i\theta} = \cos\theta + i\sin\theta$ means that $e^{i\theta}$ turns eternally around a circle of radius 1.

Why bother with the complex plane? In one direction, this simplifies the notation of Fourier analysis; instead of talking about sines and cosines we can use one coefficient for each frequency. In the other direction, it is easier to multiply complex numbers when they are represented in terms of sines and cosines, giving a geometrical interpretation of this multiplication, shown in Figure 4.

The formula

$$r_1(\cos\theta_1 + i\sin\theta_1) \times r_2(\cos\theta_2 + i\sin\theta_2)$$
$$= r_1 r_2 \Big(\cos(\theta_1 + \theta_2) + i\sin(\theta_1 + \theta_2) \Big)$$

is at the heart of trigonometry. If we carry out the multiplication of the left-hand term, we get

$$r_1 r_2 \big((\cos\theta_1 \cos\theta_2 - \sin\theta_1 \sin\theta_2) + i(\sin\theta_1 \cos\theta_2 + \cos\theta_1 \sin\theta_2) \big)$$
$$= r_1 r_2 \big(\cos(\theta_1 + \theta_2) + i\sin(\theta_1 + \theta_2) \big).$$

The Complex Plane

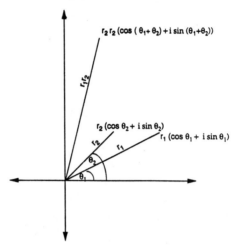

Figure 4. It is easier to multiply complex numbers when they are represented in terms of sines and cosines. The product of the complex numbers $r_1(\cos\theta_1 + i\sin\theta_1)$ and $r_2(\cos\theta_2 + i\sin\theta_2)$ is $r_1 r_2(\cos(\theta_1 + \theta_2) + i\sin(\theta_1 + \theta_2))$. That is, you multiply the lengths and add the polar angles.

Now we divide both sides by $r_1 r_2$. The two real parts, on the left and the right, must be equal, as must the two imaginary parts, so we have:

$$\cos(\theta_1 + \theta_2) = \cos\theta_1 \cos\theta_2 - \sin\theta_1 \sin\theta_2 \qquad \text{(B.1)}$$

and

$$\sin(\theta_1 + \theta_2) = \sin\theta_1 \cos\theta_2 + \cos\theta_1 \sin\theta_2, \qquad \text{(B.2)}$$

the central formulas of trigonometry.

APPENDIX C

Integrals

Dirichlet's proof of the convergence of Fourier series made it clear to mathematicians that a precise definition of integrals was needed. The first successful attempt was due to Bernhard Riemann. The underlying idea is simple. We cut the function with vertical lines, as shown in Figure 1.

For each column, we construct the largest rectangle under the graph and the smallest rectangle above it, as shown in Figure 2. If, as we space the vertical lines closer and closer together, the sum of the areas of the "lower rectangles" approaches that of the "upper rectangles," we say that the function is *Riemann integrable*, and the common limit is the integral. That integral can then be computed in different ways. The rectangles whose areas are added together can be constructed as in Figure 2, or according to some other criterion, such as by using the value of the function at the middle of the column as the height of the rectangle.

Although relatively straightforward to compute, the Riemann integral is only suited to continuous (or nearly continuous) functions. Interesting functions exist that cannot be integrated this way, for example, the function that equals 1 at all

Figure 1. The Riemann integral: the area under the graph is measured by cutting it into narrower and narrower columns and approximating each column with two rectangles, one rectangle just above the graph, the other just below. If, as the columns get narrower, the sum of the areas of the smaller rectangles approaches that of the larger rectangles, the function is *Riemann integrable*, and the common limit is the integral.

271

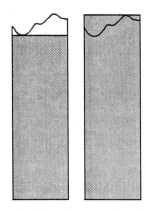

Figure 2. For each column in Figure 1 we construct the largest rectangle under the graph and the smallest rectangle above the graph.

rational numbers and 0 at all irrational numbers. The *Lebesgue integral* applies to such pathological functions (as well as to civilized ones); it is thus an essential tool for talking about quantum mechanics, for example.

For the Lebesgue integral, as for the Riemann integral, we work by discretizing a continuous situation, cutting the function into narrower and narrower intervals. But, as shown in Figure 3, we use horizontal rather than vertical lines. The picture looks similar, but the process corresponds to a completely different way of thinking. With the Riemann integral we look first at the input—the values of x—and then at the output—the values of $f(x)$. The Lebesgue integral looks first at the output—the values of $f(x)$—and then at the values of x that produced that output. It divides up the function's *domain* (the values of x) according to

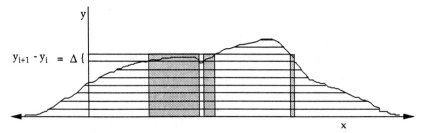

Figure 3. The Lebesgue integral looks first at the output—the values of $f(x)$—and then at the values of x that produced that output. Here the shaded zones indicate those values of x that produce an outcome $f(x)$ satisfying $y_i < f(x) \leq y_{i+1}$.

what happens in its *range* (the values of $f(x)$). This makes it possible to talk about the integral of peculiar functions like the one that equals 1 at all rational numbers and 0 at all irrational numbers.

Let us contrast what Riemann and Lebesgue say about this function. Riemann says: partition the domain $[a, b]$ into short intervals, and for each interval find the minimum and the maximum of the function. Unfortunately, the minimum is always 0 and the maximum is always 1, so the sum of the areas of the upper rectangles is the height 1 times the length $b - a$ and the sum of the areas of the lower rectangles is 0. They do not tend to each other as the partition becomes finer, so the Riemann integral does not exist.

Lebesgue says: partition the range into small intervals, and for each interval find the values of x so that $f(x)$ is in that interval. As soon as 0 and 1 are in different intervals, we will find for the interval containing 0 the set of irrationals in $[a, b]$ and for the interval containing 1 the set of rational numbers in [a,b]. The Lebesgue integral becomes

$$(0 \times \text{the length of the set of irrationals in } [a, b])$$
$$+ (1 \times \text{the length of the set of rationals in } [a, b]).$$

The question has become, "How do you measure lengths of such bizarre subsets?" This is why Lebesgue integration is often called *measure theory*.

In the above case, the irrationals in [a,b] have length (or measure) 1 and the rationals have length 0; the contribution of the rational numbers doesn't count at all, and the Lebesgue integral of this function is 0. This may seem unreasonable or even clearly false, but one can convince oneself that it is true. Consider listing all the rational numbers between 0 and 1. One way would be to take different denominators in order:

$$0, 1, \frac{1}{2}, \frac{1}{3}, \frac{2}{3}, \frac{1}{4}, \frac{3}{4}, \frac{1}{5}, \frac{2}{5}, \frac{3}{5}, \frac{4}{5}, \frac{1}{6}, \frac{5}{6}, \dots;$$

the list is of course infinite. Choose some small length ϵ ("epsilon"), and mark around 0 an interval of length $\epsilon/2$, centered at 0. Then mark one of length $\epsilon/4$ centered at 1, one of length $\epsilon/8$ centered at $1/2$, and so on. The sum of the lengths of all the marked intervals is at most

$$\frac{\epsilon}{2} + \frac{\epsilon}{4} + \frac{\epsilon}{8} + \frac{\epsilon}{16} + \dots = \epsilon,$$

but every rational number is covered by these intervals. All the rationals between 0 and 1 (and by the same reasoning in any other interval) have been put in a set of arbitrarily short length: the rational numbers thus have *measure* 0. (This can be expressed in terms of probability, since measure theory and probability theory are the same thing. If you choose a number between 0 and 1 in base 2 at random by tossing a coin and assigning 1 to heads and 0 to tails:

.111110111011000111100...

the probability is zero that you will land on a rational number.)

It may appear from this example that Lebesgue integration was just devised to deal with functions that, as Poincaré complained, "look as little as possible like decent and useful functions." But a function like the one described above is not just an interesting freak but the limit of a sequence of more "normal" functions. The function that is 1 at just a few rationals—say, the numbers $\frac{1}{2}$, $\frac{1}{3}$, $\frac{1}{4}$, $\frac{2}{3}$, $\frac{3}{4}$—and is 0 elsewhere *is* Riemann integrable; as one spaces the vertical lines closer and closer together, the rectangles that contain those numbers get skinnier and skinnier, and the sum of their areas approaches 0. As you make the function equal 1 at more and more rationals, this still holds true. The limit of this process is the function that equals 1 at all rationals, and 0 at the irrationals, which is not Riemann integrable.

"So for Riemann integration you have a sequence of integrable functions whose limit is not integrable. That makes it very hard to use Riemann integration to study limiting behavior, which is of such great interest in analysis," says Michael Frazier of Michigan State University. "The real advantage of Lebesgue integration is its stability under limiting conditions."

The Fourier Transform:
The Different Conventions

Three different equations exist for the Fourier transform, depending on where one puts the 2π. Each exists in two versions, depending on whether the minus sign is put in the exponent of the direct transform or in the exponent of the inverse transform. We list them here for those readers who may be confused on seeing different equations in different texts.

In the equations (D.1) and (D.2), ξ is measured in hertz (cycles per second) and the 2π are in the exponents. We have used the equations (D.1) in this book.

$$\hat{f}(\xi) = \int_{-\infty}^{\infty} e^{2\pi i x \xi} f(x)\, dx \quad \text{and} \quad f(x) = \int_{-\infty}^{\infty} e^{-2\pi i x \xi} \hat{f}(\xi)\, d\xi \qquad \text{(D.1)}$$

$$\hat{f}(\xi) = \int_{-\infty}^{\infty} e^{-2\pi i x \xi} f(x)\, dx \quad \text{and} \quad f(x) = \int_{-\infty}^{\infty} e^{2\pi i x \xi} \hat{f}(\xi)\, d\xi \qquad \text{(D.2)}$$

In the equations (D.3)–(D.6), the variable of frequency, ξ, is measured in radians per second (1 radian $= \frac{1}{2\pi}$ of a circle). In the equations (D.3) and (D.4), the 2π are in the inverse transform. In (D.5) and (D.6), the 2π are shared by the direct transform and the inverse transform.

$$\hat{f}(\xi) = \int_{-\infty}^{\infty} e^{i x \xi} f(x)\, dx \quad \text{and} \quad f(x) = \frac{1}{2\pi} \int_{-\infty}^{\infty} e^{-i x \xi} \hat{f}(\xi)\, d\xi \qquad \text{(D.3)}$$

$$\hat{f}(\xi) = \int_{-\infty}^{\infty} e^{-i x \xi} f(x)\, dx \quad \text{and} \quad f(x) = \frac{1}{2\pi} \int_{-\infty}^{\infty} e^{i x \xi} \hat{f}(\xi)\, d\xi \qquad \text{(D.4)}$$

$$\hat{f}(\xi) = \frac{1}{\sqrt{2\pi}} \int_{-\infty}^{\infty} e^{i x \xi} f(x)\, dx \quad \text{and} \quad f(x) = \frac{1}{\sqrt{2\pi}} \int_{-\infty}^{\infty} e^{-i x \xi} \hat{f}(\xi)\, d\xi \qquad \text{(D.5)}$$

$$\hat{f}(\xi) = \frac{1}{\sqrt{2\pi}} \int_{-\infty}^{\infty} e^{-i x \xi} f(x)\, dx \quad \text{and} \quad f(x) = \frac{1}{\sqrt{2\pi}} \int_{-\infty}^{\infty} e^{i x \xi} \hat{f}(\xi)\, d\xi \qquad \text{(D.6)}$$

APPENDIX E

The Sampling Theorem: A Proof

The sampling theorem says that if the range of frequencies of a signal $f(t)$ is M hertz (cycles per second), then one can reconstruct the signal exactly by sampling it $2M$ times a second.

Let us choose our units of time (or frequency) so that the bandwidth of our signal is $[-\frac{1}{2}, \frac{1}{2}]$: that is, its Fourier transform $\hat{f}(\tau)$ has nonzero values only for $|\tau| \le \frac{1}{2}$, as shown in Figure 1. Let g be the periodic function of period 1 that coincides with $\hat{f}$ for $|\tau| \le \frac{1}{2}$.

We will prove that the Fourier coefficients of g are also the values $f(n)$ of f at the integers $n = \ldots -2, -1, 0, 1, 2 \ldots$. If we know the values of f at the integers, we know g which gives us $\hat{f}$. Inverting the Fourier transform, we can reconstruct the function f.

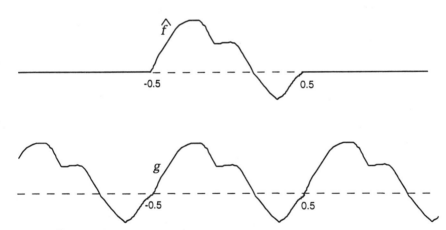

Figure 1. The Fourier transform $\hat{f}$, shown above, has nonzero values only for $|\tau| \le \frac{1}{2}$ (i.e., the bandwidth of the signal f is $[-\frac{1}{2}, \frac{1}{2}]$). The function g is the periodic function of period 1 that coincides with $\hat{f}$ for $|\tau| \le \frac{1}{2}$.

Let us write g as a Fourier series:

$$g(\tau) = \sum_{n=-\infty}^{\infty} c_n e^{-2\pi i n \tau}; \tag{E.1}$$

the coefficients of this series are given by

$$c_{-n} = \underbrace{\int_{-\frac{1}{2}}^{\frac{1}{2}} g(\tau) e^{-2\pi i n \tau} d\tau}_{[1]} = \underbrace{\int_{-\frac{1}{2}}^{\frac{1}{2}} \hat{f}(\tau) e^{-2\pi i n \tau} d\tau}_{[2]} = \underbrace{\int_{-\infty}^{\infty} \hat{f}(\tau) e^{-2\pi i n \tau} d\tau}_{[3]} = f(n).$$

$$\tag{E.2}$$

If this seems confusing, remember that g is periodic and has a Fourier *series*, not transform; it has Fourier coefficients only for those frequencies that are whole number multiples of the base frequency. Since g coincides with $\hat{f}$ for $|\tau| \leq \frac{1}{2}$, we can substitute $\hat{f}(\tau)$ for $g(\tau)$ in [2]. Since $\hat{f}$ has nonzero values only for $|\tau| \leq \frac{1}{2}$, [2] equals [3]. By the theorem of Parseval, the inverse Fourier transform then gives us the values $f(n)$.

We have finished. If we know the values of f at the integers—measuring the signal once per unit time, or twice its bandwidth—we can reconstruct f. However, the process above is somewhat indirect. Below we give a more straightforward reconstruction formula.

We begin by writing f as the inverse Fourier transform of $\hat{f}$:

$$f(t) = \int_{-\infty}^{\infty} \hat{f}(\tau) e^{-2\pi i \tau t} d\tau.$$

Since $\hat{f}$ has nonzero values only for $|\tau| \leq \frac{1}{2}$, we see that

$$f(t) = \int_{-\frac{1}{2}}^{\frac{1}{2}} \hat{f}(\tau) e^{-2\pi i \tau t} d\tau.$$

In the interval $[-1/2, 1/2]$, the function $\hat{f}$ is given by the formula (E.1), so we substitute the sum in (1) for $\hat{f}(\tau)$:

$$f(t) = \int_{-\frac{1}{2}}^{\frac{1}{2}} \left(\sum_{n=-\infty}^{\infty} c_{-n} e^{2\pi i n \tau} e^{-2\pi i \tau t} \right) d\tau.$$

(We moved the minus sign from the superscript to the subscript, which we can do, since we are considering all n from $-\infty$ to ∞.)

Now, using (E.2) above, we replace c_{-n} by $f(n)$ and we exchange the sum and the integral. (Such an exchange isn't always justified but is here.)

$$f(t) = \sum_{n=-\infty}^{\infty} \left(\int_{-\frac{1}{2}}^{\frac{1}{2}} f(n) \, e^{2\pi i \tau (n-t)} \right) d\tau = \sum_{n=-\infty}^{\infty} \left(f(n) \int_{-\frac{1}{2}}^{\frac{1}{2}} e^{2\pi i \tau (n-t)} \right) d\tau.$$

So we can reconstruct f from its values at the integers. We can go one step further and compute the integral:

$$\int_{-\frac{1}{2}}^{\frac{1}{2}} e^{2\pi i \tau (n-t)} d\tau = \left[\frac{1}{2\pi i (n-t)} e^{2\pi i \tau (n-t)} \right]_{-\frac{1}{2}}^{\frac{1}{2}} = \frac{\sin \pi (n-t)}{\pi (n-t)},$$

which gives the reconstruction formula:

$$f(t) = \sum_{n=-\infty}^{\infty} f(n) \frac{\sin \pi (n-t)}{\pi (n-t)}.$$

The Heisenberg Uncertainty Principle: A Proof

Let f be a function of the real variable x such that

$$\int_{-\infty}^{\infty} |f(x)|^2 dx = 1.$$

We will think of $|f|^2$ as a probability density; $|\hat{f}|^2$ is then also a probability density. The Heisenberg uncertainty principle says that the product of the variance of x for $|f|^2$ and the variance of ξ for $|\hat{f}|^2$ is at least $\frac{1}{16\pi^2}$:

$$\underbrace{\left(\int_{-\infty}^{\infty} (x - x_m)^2 |f(x)|^2 dx \right)}_{\text{variance of } x} \underbrace{\left(\int_{-\infty}^{\infty} (\xi - \xi_m)^2 |\hat{f}(\xi)|^2 d\xi \right)}_{\text{variance of } \xi} \geq \frac{1}{16\pi^2}. \quad \text{(F.1)}$$

(The exact number to the right of this inequality depends on the formula used for the Fourier transform.)

If we set

$$g(x) = e^{2\pi i x \xi_m} f(x + x_m),$$

then g is also normalized, $\hat{g}$ is given by the equation

$$\hat{g}(\xi) = e^{-2\pi i x_m (\xi + \xi_m)} \hat{f}(\xi + \xi_m),$$

and if we substitute those values for g and $\hat{g}$ in the following formula, after a bit of computation we get

$$\left(\int_{-\infty}^{\infty} x^2 |g(x)|^2 dx \right) \left(\int_{-\infty}^{\infty} \xi^2 |\hat{g}(\xi)|^2 d\xi \right)$$
$$= \left(\int_{-\infty}^{\infty} (x - x_m)^2 |f(x)|^2 dx \right) \left(\int_{-\infty}^{\infty} (\xi - \xi_m)^2 |\hat{f}(\xi)|^2 d\xi \right).$$

So if we prove

$$\left(\int_{-\infty}^{\infty} x^2 |f(x)|^2 dx \right) \left(\int_{-\infty}^{\infty} \xi^2 |\hat{f}(\xi)|^2 d\xi \right) \geq \frac{1}{16\pi^2} \qquad \text{(F.2)}$$

for any normalized function f, we will have proved it in particular for g, proving that formula (F.1) is correct.

For the first step of the proof we will use the fact that

$$\xi \hat{f}(\xi) = -\frac{1}{2\pi i} \widehat{f'}(\xi).$$

That is, multiplying by ξ in Fourier space corresponds (to within a factor of $-\frac{1}{2\pi i}$) to differentiating in position space. (It may seem puzzling that there are "hats" on both sides of this equality. Think of the analogous statement that adding an "s" in English, to form a plural, is equivalent in Latin—restricting ourselves to masculine words—to changing "us" to "i." The English-Latin dictionary is equivalent to the Fourier transform. Adding an "s" in English and then looking up the Latin in the dictionary gives the same result as first using the dictionary, then changing "us" to "i." Similarly, differentiating f to get f' and then using the "dictionary" or Fourier transform to get $\widehat{f'}$ gives the same result as first using the dictionary to get $\hat{f}$ and then multiplying by ξ.)

Often this relationship is used in order to avoid differentiating: one goes temporarily into Fourier space, where computations are easier. Here we do the reverse, replacing the $\xi^2 |\hat{f}(\xi)|^2$ in (F.2) by $\frac{1}{4\pi^2} |\widehat{f'}(\xi)|^2$, to obtain:

$$\frac{1}{4\pi^2} \left(\int_{-\infty}^{\infty} x^2 |f(x)|^2 dx \right) \left(\int_{-\infty}^{\infty} |\widehat{f'}(\xi)|^2 d\xi \right) \geq ? \qquad \text{(F.3)}$$

Now we get rid of the "hat" in the second term, using the fact (from the theorem of Parseval) that the Fourier transform of a function is an *isometry*: the square of the length of $\hat{f}$ is equal to the square of the length of f. Thus we can replace $|\widehat{f'}(\xi)|^2$ by $|f'(x)|^2$ to get:

$$\frac{1}{4\pi^2} \left(\int_{-\infty}^{\infty} x^2 |f(x)|^2 dx \right) \left(\int_{-\infty}^{\infty} |f'(x)|^2 dx \right) \geq ? \qquad \text{(F.4)}$$

The next step is the most important, since it allows us to complete the inequality. We will use the Schwarz inequality, which concerns scalar products. As discussed in *Orthogonality and Scalar Products*, p. 153, the scalar product of two vectors can be written either $\cos \theta \, |\vec{v}||\vec{w}|$ or $< \vec{v}, \vec{w} >$, where θ is the angle between the vectors $\vec{v}$ and $\vec{w}$; since $\cos \theta \leq 1$, then

$$|\vec{v}|^2 |\vec{w}|^2 \geq \quad |< \vec{v}, \vec{w} >|^2.$$

We saw in the same article that a scalar product can be seen as an integral:

$$|<\vec{v},\vec{w}>| = \int_{-\infty}^{\infty} \vec{v}(x)\vec{w}(x)\,dx.$$

Setting $\vec{v}(x) = x|f(x)|$ and $\vec{w}(x) = |f'(x)|$, we have

$$\frac{1}{4\pi^2}\underbrace{\left(\int_{-\infty}^{\infty} x^2|f(x)|^2 dx\right)}_{|\vec{v}|^2}\underbrace{\left(\int_{-\infty}^{\infty}|f'(x)|^2 dx\right)}_{|\vec{w}|^2} \geq \frac{1}{4\pi^2}\underbrace{\left(\int_{-\infty}^{\infty}|xf(x)f'(x)|\,dx\right)^2}_{|<\vec{v},\vec{w}>|^2}.$$

(F.5)

For any two complex numbers a and b

$$|ab| \geq \frac{1}{2}(a\bar{b} + \bar{a}b),$$

where "bar" indicates a complex conjugate: $\bar{a}$ is the complex conjugate of a. This gives us

$$\cdots \geq \frac{1}{4\pi^2}\left(\frac{1}{2}\int_{-\infty}^{\infty}\left(\underbrace{xf(x)}_{a}\,\underbrace{\overline{f'(x)}}_{\bar{b}} + \underbrace{x\overline{f(x)}}_{\bar{a}}\,\underbrace{f'(x)}_{b}\right)dx\right)^2. \qquad \text{(F.6)}$$

Since

$$\frac{d}{dx}|f(x)|^2 = f(x)\overline{f'(x)} + f'(x)\overline{f(x)},$$

we get

$$\cdots \geq \frac{1}{16\pi^2}\left(\int_{-\infty}^{\infty} x\frac{d}{dx}|f(x)|^2 dx\right)^2 \qquad \text{(F.7)}$$

This becomes (by integration by parts):

$$\cdots \geq \frac{1}{16\pi^2}\left(-\int_{-\infty}^{\infty}|f(x)|^2 dx\right)^2. \qquad \text{(F.8)}$$

By definition, the integral in (F.8) equals 1, and fortunately the minus sign disappears when the integral is squared.

$$\left(\int_{-\infty}^{\infty} x^2|f(x)|^2\,dx\right)\left(\int_{-\infty}^{\infty}\xi^2|\hat{f}(\xi)|^2\,d\xi\right) \geq \frac{1}{16\pi^2},$$

which gives us

$$\left(\int_{-\infty}^{\infty}(x - x_m)^2|f(x)|^2 dx\right)\left(\int_{-\infty}^{\infty}(\xi - \xi_m)^2|\hat{f}(\xi)|^2 d\xi\right) \geq \frac{1}{16\pi^2}.$$

The Fourier Transform of a Periodic Function

The purpose of this appendix is to sketch an explanation of why the periodic function A described in *Multiresolution*, page 165, whose Fourier series is

$$A(\xi) = \sum_{n=-\infty}^{\infty} a_n \, e^{2\pi i n \xi},$$

is the Fourier transform of the low-pass digital filter a consisting of the numbers a_n. More generally this will explain why we can consider the *sum* of a Fourier series the *Fourier transform* of its coefficients.

A priori the words "the Fourier transform of a sequence of numbers" have no meaning; *functions* have Fourier transforms, but how can a sequence of numbers have a Fourier transform?

We will show that summing a Fourier series is a special case of the Fourier transform. That is, we will show the sense in which a periodic function like A has a Fourier transform, which consists of the coefficients of its Fourier series (the sequence a). Since the inverse Fourier transform is the same as the Fourier transform (with a change of sign, as shown by the formulas p. 275), we will then be able to say, interchangeably, that A is the Fourier transform of a or that a is the Fourier transform of A.

But how can a periodic function like A have a Fourier transform, as opposed to a Fourier series? The formula for the Fourier transform of a function that decreases fast enough at infinity so that the area under its graph is finite is

$$\hat{f}(\xi) = \int_{-\infty}^{\infty} f(x) \, e^{2\pi i x \xi} dx. \tag{G.1}$$

That is, we compute the integral of the product of the function and the complex exponential of each frequency ξ. For a periodic function those integrals aren't finite, so it would appear that formula (G.1) has no meaning when f is periodic.

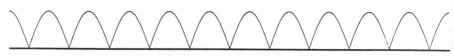

Figure 1. The periodic function $f(x) = |\sin \pi x|$. It isn't obvious in what sense such a function can have a Fourier transform. To compute a Fourier transform we compute the integral of the product of the function and the complex exponential of each frequency ξ. For a periodic function those integrals aren't finite.

To see how we can attach some meaning to the notion of a Fourier transform of a periodic function, let's look at a particular periodic function, for example $f(x) = |\sin \pi x|$, whose graph is shown in Figure 1.

As we said, we can't use formula (G.1) to compute its Fourier transform. But we can cheat by computing (by computer) the Fourier transform of a *truncated* version of this function, which *will* have a Fourier transform, for example, the function $g_m(x)$ that equals $|\sin \pi x|$ in the interval $[-m, m]$ and 0 outside that interval. We do this by computing the integral of the product $(f(x) \cos 2\pi x \xi)$ only between $x = -m$ and $x = m$, getting the Fourier transform:

$$\hat{g}_m(\xi) = \int_{-m}^{m} |\sin \pi x| \cos 2\pi x \xi \, dx.$$

For $m = 10$ and for $-5.5 \leq \xi \leq 5.5$, we get the graph shown in Figure 2. (The sines don't contribute to this integral.)

A true Fourier series has coefficients a_n only at the integers; the function $\hat{g}_m(\xi)$ has spikes at the integers and oscillates rapidly between the integers. When m gets larger, the function g_m repeats for a longer time and looks more like our periodic function f; the oscillations of $\hat{g}_m(\xi)$ between the integers become more and more rapid, and the spikes at the integers become higher and higher.

The rapid oscillations cancel each other, and as m tends to infinity the functions $\hat{g}_m(\xi)$ *tend weakly*[1] to a sum of point masses at the integers and 0

---

[1] A sequence of functions f_k approaches a function f *weakly* if for any smooth function g with compact support, we have

$$\lim_{k \to \infty} \int f_k(x)g(x) \, dx = \int f(x)g(x) \, dx.$$

For instance, $f_k(x) = \sin kx$ approaches $f(x) = 0$ weakly. This is why for high frequencies, the Fourier coefficients of smooth functions are small, as mentioned in *Computing Fourier Coefficients with Integrals*.

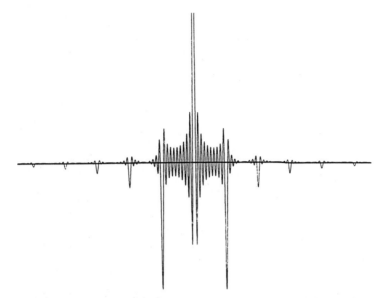

Figure 2. The Fourier transform of the function that is $|\sin \pi x|$ in the interval $[-10, 10]$ and 0 outside that interval (i.e., a truncated version of the periodic function shown in Figure 1). The transform was computed for frequencies $-5.5 \leq \xi \leq 5.5$. As the function is truncated further and further from the origin—approaching the periodic function $|\sin \pi x|$—the corresponding Fourier transform tends weakly to a sum of point masses at the integers, and 0 between the integers.

between the integers. More precisely, they approach (weakly) the distribution:

$$\hat{f}(\xi) = \sum_{k=-\infty}^{\infty} a_k \delta(\xi - k), \tag{G.2}$$

in which δ is the Dirac δ-function ("delta function"). This δ-function is not a function at all but the distribution that represents an abstract "unit mass" at 0, so that $\delta(\xi - k)$ (i.e., the δ-function translated by k) is the unit mass at frequency k. For each integer k, the distribution $\delta(\xi - k)$ is weighted by the a_k, the coefficients of the function f's Fourier series. (The a_k give the units of "mass." For example, if f is an audio signal, its units measure pressure, and $a_k \delta(\xi - k)$ is the pressure contributed by the frequency k.)

If we Fourier transform $\hat{f}(\xi)$, Parseval's theorem says that we get $f(-x)$. Fourier's statement (proved by Dirichlet) that a periodic function *is* the sum of its Fourier series is a special case of Parseval's theorem. The Fourier transform of $\hat{f}(\xi)$ is:

$$\hat{\hat{f}}(-x) = \int_{-\infty}^{\infty} \sum_{k=-\infty}^{\infty} a_k \delta(\xi - k)\, e^{-2\pi i \xi x}\, d\xi. \tag{G.3}$$

Exchanging the sum and integral gives:

$$\hat{\hat{f}}(-x) = \sum_{k=-\infty}^{\infty} a_k \int_{-\infty}^{\infty} \delta(\xi - k)\, e^{-2\pi i \xi x}\, d\xi. \tag{G.4}$$

This looks complicated but it isn't. One operation involving δ-functions is very easy. Integrating the product of a function g and a δ-function translated by k simply evaluates the function g at k:

$$\int_{-\infty}^{\infty} \delta(\xi - k)\, g(\xi)\, d\xi = g(k).$$

So we can simplify the integral in (G.4):

$$\int_{-\infty}^{\infty} \delta(\xi - k)\, e^{-2\pi i \xi x}\, d\xi = e^{-2\pi i k x},$$

which gives

$$\hat{\hat{f}}(-x) = \sum_{k=-\infty}^{\infty} a_k\, e^{-2\pi i k x} = f(x).$$

So the inverse Fourier transform of the sequence a_k (more precisely, of $\sum_{k=-\infty}^{\infty} a_k \delta(\xi-k)$) consists of using the a_k as the coefficients of a Fourier series and summing the series. In particular, the function A described in *Multiresolution* is the Fourier transform of the low-pass filter consisting of the numbers a_n (or the filter's inverse Fourier transform, depending on a change of sign).

The Fourier Series of a Function that Decreases at Infinity

We have not gone about this in the usual order. Since students generally learn about Fourier series before learning about Fourier transforms, the problem that usually arises is the reverse. How can we apply the idea of a Fourier series to a nonperiodic function that decreases fast enough at infinity so that its integral is finite?

To do this we make a periodic approximation of our nonperiodic function. We cut off the ends of the function at $-T/2$ and $T/2$ for some large

T and we make a periodic function by stringing together copies of the truncated function. We then look at its Fourier series. The function is periodic of period T, so its Fourier series consists of coefficients for the frequencies at $\ldots, -\frac{2}{T}, -\frac{1}{T}, 0, \frac{1}{T}, \frac{2}{T}, \ldots$, instead of at the integers or at multiples of 2π.

As T becomes longer, the coefficients are spaced closer and closer together. When $T = \infty$, our periodic function has become nonperiodic, and the discrete Fourier series has become a continuous Fourier transform; we are looking at all possible frequencies. (Of course, in practice a Fourier transform is computed for some sampling of frequencies, as described in *The Fast Fourier Transform*, p. 141.)

An Example of an Orthonormal Basis and a Proof of Fourier's Result

The classical example of an orthonormal basis is the basis formed by the trigonometric functions $e^{2\pi inx}$ (where n is a whole number), in the space of square integrable functions that are periodic of period 1. Decomposing a function in this basis means writing it as a Fourier series.

It is easy to show that these functions are orthonormal. Saying that the functions $e^{2\pi inx}$ (which we will call e_n) are orthogonal means that the scalar product $< e_n, e_m >$ is zero when $n \neq m$. Saying that they are ortho*normal* means that this scalar product is 1 when $n = m$. (The length of a vector is the square root of the scalar product of a vector with itself.)

The scalar product $< e_n, e_m >$ can be expressed as an integral, using complex conjugates, since e_n and e_m are complex functions:

$$< e_n, e_m >= \int_0^1 e^{2\pi inx}\overline{e^{2\pi imx}}\, dx. \qquad \text{(H.1)}$$

Now we use the fact that

$$\overline{e^{2\pi imx}} = e^{-2\pi imx}$$

to express the integral in (H.1) as

$$\int_0^1 e^{2\pi ix(n-m)}\, dx. \qquad \text{(H.2)}$$

Set $k = (n-m)$ and consider, rather than the integral in (H.2), the equivalent integral:

$$\int_0^1 (\cos 2\pi kx + i \sin 2\pi kx)\, dx. \qquad \text{(H.3)}$$

291

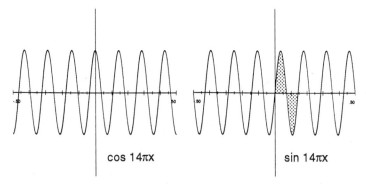

Figure 1. As shown by the above example, every function cos $2\pi kx$ and sin $2\pi kx$, where k is an integer, has zero integral over any interval of length 1: each negative part exactly compensates each positive part.

When $n \neq m$, k is a non-zero whole number, and the two functions oscillate precisely k times between 0 and 1. Their oscillations compensate exactly, giving an integral of zero, as shown in Figure 1. So our functions are orthogonal.

When $n = m$, then $k = 0$, and we integrate, between 0 and 1, the constant function $\cos 0 + i \sin 0 = 1$. This integral equals 1; our functions thus have length 1.

(Those who find this proof by pictures too easy can get the same result by computing the integral in (H.1).)

We've not finished. Now we must show that these orthonormal functions form a basis in the space of square summable functions that are periodic of period 1. That is, that they can represent exactly any function in this space (or, equivalently, that a finite number of these functions can approximate the function with arbitrary precision). We will prove it for any continuous function.

Let us consider the function $g(x) = 1 + \cos 2\pi x$, between $-\frac{1}{2}$ and $\frac{1}{2}$, shown at left in Figure 2: always positive, and with the value 2 at 0. Raise this function to a large power N, say $N = 10\,000$, to make it into a giant. Next force it to have integral 1 by multiplying it by the appropriate constant (in this case, by $\frac{2^N (N!)^2}{(2N)!}$). The function g_N that we get has an unimaginably high peak at 0, and equals almost 0 everywhere else, as shown at right in Figure 2; it is a very good approximation to a "delta function," which has infinite height, no width and integral 1.

We want to show that this emaciated giant is a finite sum of trigonometric functions (cosines in this case), and that it allows us to approximate any continuous function f. Let's consider the second question first. We will integrate the

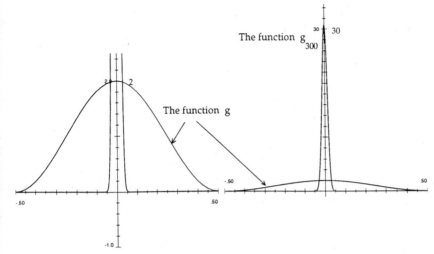

Figure 2. At left we see the function $g(x) = 1 + \cos 2\pi x$ and the bottom part of g raised to the power $N = 300$. At right we change the scale of the g-axis to give a better view of both functions. As $N \to \infty$, the function g raised to the Nth power can be used to approximate any continuous function f more and more precisely, as shown in equation (H.4). Since g raised to the Nth power is a sum of trigonometric functions, we see that the trigonometric functions form a basis.

product $g_N f$. This integral is almost $f(0)$:

$$\int_{-\frac{1}{2}}^{\frac{1}{2}} f(x) g_N(x)\, dx \approx \int_{-\frac{1}{2}}^{\frac{1}{2}} f(0) g_N(x)\, dx = f(0) \underbrace{\int_{-\frac{1}{2}}^{\frac{1}{2}} g_N(x)\, dx}_{\text{1 by definition}} = f(0)$$

The first integral is almost equal to the second because g_N is almost zero except very near 0, making the product $g_N f$ to be almost 0 also, except very near 0. The approximation becomes more and more exact as N tends to infinity.

We are interested in all values of f, not just $f(0)$. So we translate g_N by y, creating a new function $g_N(x - y)$, and use it to create $f_N(y)$, which is (almost) equal to $f(y)$:

$$f_N(y) = \underbrace{\int_{(-\frac{1}{2}+y)}^{(\frac{1}{2}+y)} f(x) g_N(x - y)\, dx}_{\text{trigonometric polynomial}} \approx \int_{(-\frac{1}{2}+y)}^{(\frac{1}{2}+y)} f(y) g_N(x - y)\, dx = f(y).$$

$$(\text{H.4})$$

We will now show that $f_N(y)$ is a trigonometric polynomial: the sum of a finite number of trigonometric functions.

Let's return to our original function g: $g_1 = 1 + \cos 2\pi x$. Using the equation (which follows from the Euler formula)

$$\cos \alpha = \frac{e^{i\alpha} + e^{-i\alpha}}{2},$$

we have

$$1 + \cos 2\pi x = \frac{e^{-2\pi i x}}{2} + 1 + \frac{e^{2\pi i x}}{2}.$$

If we raise the right-hand term to the power N we get a sum of $2N + 1$ trigonometric functions beginning with $\left(\frac{e^{-2\pi i x}}{2}\right)^N = \frac{1}{2^N} e^{-2\pi i N x}$ and ending with $\left(\frac{e^{2\pi i x}}{2}\right)^N = \frac{1}{2^N} e^{2\pi i N x}$.

So we can write

$$g_N(x) = \sum_{n=-N}^{N} \alpha_n e^{2\pi i n x}.$$

Substituting this sum—with $(x - y)$ rather than x in the exponent—for g_N in formula (H.4), we get:

$$f_N(y) = \int_{-\frac{1}{2}}^{\frac{1}{2}} f(x) \sum_{n=-N}^{N} \alpha_n e^{2\pi i n(x-y)} dx.$$

Exchanging the integral and the sum then gives us:

$$f(y) \approx f_N(y) = \sum_{n=-N}^{N} \alpha_n e^{-2\pi i n y} \underbrace{\int_{-\frac{1}{2}}^{\frac{1}{2}} e^{2\pi i n x} f(x) dx}_{\text{the coefficient } c_n} = \sum_{n=-N}^{N} c_n \alpha_n e^{-2\pi i n y}.$$

Any continuous, periodic function can thus be uniformly approximated by a finite sum of trigonometric functions. This proof can easily be extended to functions with step discontinuities, using a somewhat weaker definition of approximation. Square summable periodic functions with nastier discontinuities—even the one that jumps incessantly from plus infinity to minus infinity, discussed in *Traveling from One Function Space to Another*—can also be approximated, using a yet weaker definition of approximation.

Further Reading

Fourier Biographies

Joseph Fourier, 1768–1830, I. Gratten-Guinness and J.R. Ravetz, MIT Press, Cambridge, MA, 1972.

Joseph Fourier—the Man and the Physicist, J. Herivel, Clarendon Press, Oxford, 1975.

Books about Fourier Analysis and Fourier Transforms

The DFT: An Owner's Manual for the Discrete Fourier Transform, William L. Briggs and Van Emden Henson, SIAM, Philadelphia, 1995.

Fourier Analysis, T. W. Körner, Cambridge University Press, Cambridge, New York, 1988.

(These two books are a small sample of the extensive literature in the field.)

Books about Wavelets

Multiresolution Signal Decomposition Transforms, Subbands, and Wavelets, A.N. Akansu and R.A. Haddad, Academic Press, Inc., Boston, 1992. (Intended for "graduate students who have a working knowledge of linear system theory and Fourier analysis, some linear algebra, random signals and processes, and an introductory course in digital signal processing.")

Wavelet, Subband and Block Transforms in Communications and Multimedia, A.N. Akansu and M. Medley, eds., Kluwer Academic Publishers, Norwood, MA, 1998.

Subband and Wavelet Transforms: Design and Applications, A.N. Akansu and M.J.T. Smith, eds., Kluwer Academic Publishers, Norwood, MA, 1996.

Wavelets in Medicine and Biology, A. Aldroubi and M. Unser, eds., CRC Press, Boca Raton, FL, 1996. (For information on wavelets and medicine see also [Unser, Aldroubi] and Metin Akay, "Wavelet Applications in Medicine," *IEEE Spectrum*, Vol. 34, No 5, pp. 50–56 May, 1997.)

Wavelets: Mathematics and Applications, John J. Benedetto and Michael W. Frazier (eds.), CRC Press, Boca Raton, FL, 1993. (A collection of articles described by the editors as "at once mathematically precise and, for the most part, accessible to a general scientific and engineering audience," with sections on basic wavelet theory, wavelets and signal processing, and wavelets and partial differential operators.)

Introduction to Wavelets and Wavelet Transforms: A Primer, C. Sidney Burrus, Ramesh A. Gopinath, and Haitao Guo, Prentice Hall, Upper Saddle River, N.J., 1998. (Presents wavelets from the point of view of signal expansion rather than filter banks. Intended for advanced undergraduates, graduates, and people in industry; requires background only in Fourier methods and linear algebra.)

Wavelet Basics, Y.T. Chan, Kluwer Academic Publishers, Norwood, MA, 1995.

An Introduction to Wavelets, C.K. Chui, Academic Press, New York, 1992.

Wavelets: A Tutorial in Theory and Application, C.K. Chui (ed.), Academic Press, New York, 1992.

Wavelets and Multiscale Signal Processing, Albert Cohen and Robert Ryan, Chapman & Hall, London, 1995.

Ten Lectures on Wavelets, Ingrid Daubechies, SIAM, Philadelphia, PA, 1992. (Covers many aspects of wavelet theory, in continuous as well as discrete time. The first few chapters make connections with quantum physics, time–frequency localization in signal analysis, and approximation theory; the later chapters focus on orthonormal wavelet bases and how to construct wavelet bases with various properties. The book won the AMS Leroy Steele prize for exposition.)

Different Perspectives on Wavelets, Ingrid Daubechies, ed., Proc. Sympos. Appl. Math., Vol. 47, Amer. Math. Soc., Providence, RI, 1993. (Survey lectures given by Ingrid Daubechies, Yves Meyer, Pierre Gilles Lemarié-Rieusset, Philippe Tchamitchian, Gregory Beylkin, Ronald R. Coifman, M. Victor Wickerhauser, and David L. Donoho at an American Mathematical Society Short Course on Wavelets and Applications held in San Antonio, Jan. 1993.)

Wavelets and Singular Integrals on Curves and Surfaces, Guy David, Lecture Notes Math. 1465, Springer-Verlag, Berlin, 1991.

Wavelets, Fractals, and Fourier Transforms, Marie Farge, J.C.R. Hunt and J.C. Vassilicos, eds., Clarendon Press, Oxford, 1993. (Based on the proceedings of a conference held in Cambridge, December 1990.)

An Introduction to Wavelets Through Linear Algebra, Michael Frazier, Springer-Verlag, New York, expected publication 1998. (In the series Undergraduate Texts In Mathematics, this is an introduction to wavelets at the advanced undergraduate or beginning graduate level, beginning with a review of linear algebra. Wavelets are presented first in the finite dimensional context, then on the integers, then on the real line; necessary background is developed as needed. The construction and implementation of Daubechies's wavelets are described. Applications to image compression and to numerical differential equations are briefly discussed.)

A First Course on Wavelets, Eugenio Hernandez and Guido Weiss, CRC Press, Inc., Boca Raton, FL, 1996. (In the Studies in Advanced Mathematics Series edited by Steven G Krantz.)

Wavelets, An Analysis Tool, Matthias Holschneider, Oxford University Press, Oxford, 1995. (Intended for graduate students in mathematics and physics. Its main focus is the continuous wavelet transform and its applications, but it treats multiresolution analysis as well. It also has separate chapters on the analysis of fractals, and on the relations of wavelets with group theory and with functional analysis.)

Fourier Series and Wavelets, J.P. Kahane and P.G. Lemarié-Rieusset, Gordon & Breach, London, 1995. (The first section, by Kahane, discusses the history of Fourier analysis from Fourier to the present. The second, by Lemarié-Rieusset, discusses the mathematics of wavelets and wavelet theory.)

A Friendly Guide to Wavelets, G. Kaiser, Birkhäuser, Boston, 1994. (Intended as a textbook "for a one-semester introductory course on wavelet analysis aimed at graduate students or advanced undergraduates in science, engineering, and mathematics," with "emphasis on motivation and explanation rather than mathematical rigor.")

Les Ondelettes en 1989, P.G. Lemarié, ed., Lectures Notes Math. 1438, Springer-Verlag, Berlin, 1990.

A Wavelet Tour of Signal Processing, Stéphane Mallat, Academic Press, Boston, 1998. (Graduate level, intended for both mathematicians and engineers.)

Wavelets and Operators, Yves Meyer, Cambridge University Press, Cambridge, 1992. (A translation, by D. H. Salinger, of *Ondelettes et Opérateurs*, Hermann, Paris, 1990.)

Wavelets Algorithms & Applications, Yves Meyer, SIAM, Philadelphia, 1993. (A translation, by Robert Ryan, of *Les Ondelettes Algorithmes et Applications*, Armond Colin, Paris, 1992.)

Random Vibrations, Spectral and Wavelet Analysis, 3rd edition, D. E. Newland, Addison Wesley Longman, Reading, Massachusetts, 1993

Essential Wavelets for Statistical Applications and Data Analysis, R. Todd Ogden, Birkhäuser, Boston, 1997.

Joint Time-Frequency Analysis, Shie Qian and Dapang Chen, Prentice-Hall, Upper Saddle River, NJ, 1996. (Graduate textbook, with emphasis on Gabor and Wigner's time-frequency analysis; includes one chapter on wavelets.)

Wavelets and Their Applications, B. Ruskai, ed., Jones and Bartlett, Boston, 1992. (A collection of articles concerning signal analysis, numerical analysis, other applications, and theoretical developments; many originated as invited lectures at a 1990 conference that also gave rise to Daubechies's *Ten Lectures on Wavelets*.)

Wavelets for Computer Graphics: Theory and Applications, Eric J. Stollnitz, Tony D. DeRose, and David H. Salesin, Morgan Kaufmann, San Francisco, 1996. (Focuses on a generalized theory that accommodates the kinds of objects that commonly arise in computer graphics, including images, open curves, and surfaces of arbitrary topology.)

Wavelets and Filter Banks, Gilbert Strang and Truong Nguyen, Wellesley-Cambridge Press, Wellesley, MA, 1996. (Textbook for engineers, scientists, and applied mathematicians. In discrete time the text describes the theory of filter banks and perfect reconstruction. In continuous time it explains the construction and properties of wavelets.)

Analyse continue par ondelettes, Bruno Torrésani, Savoirs Actuels, InterÉditions/CNRS Éditions, Paris, 1995.

Wavelet Theory and Its Applications, Randy K. Young, Kluwer Academic Publishers, Boston, 1993. (Requires college senior-level mathematics; primarily addresses signal processing applications.)

Wavelets in Physics, J.C. van den Berg, ed., Cambridge University Press, Cambridge, to appear 1998. (Treats applications in astronomy, turbulence, plasma physics, meteorology, atomic and solid state physics, multifractal analysis, medicine and physiology, and mathematical physics. Includes chapters by J.-P. Antoine, A. Bijaoui, M. Farge, L. Hudgins, B. van Milligen, A. Fournier, A. Arnéodo, and others.)

Wavelets and Subband Coding, Martin Vetterli and J. Kovacevic, Prentice-Hall, Upper Saddle River, NJ, 1995. (First-year graduate textbook for engineers and applied mathematicians. Covers the theory and applications of wavelets and filter banks from a signal processing perspective. Prerequisite: a basic knowledge of signal processing and Fourier analysis.)

Adapted Wavelet Analysis from Theory to Software, M. Victor Wickerhauser, A K Peters, Ltd., Wellesley, MA, 1994. (Described by the publisher as "a detail-oriented text ... intended for engineers and applied mathematicians who must write computer programs to perform wavelet and related analysis on real data." Optional 3-1/2″ disk.)

A Mathematical Introduction to Wavelets, P. Wojtaszczyk, (Vol. 37 in the series London Mathematical Society Students Texts), 1997. (Aimed at senior undergraduate or beginning graduate students in mathematics. The book presents the theory of orthogonal wavelets in one and several variables and their application to analysis of important function spaces.)

Articles Aimed at Undergraduates

"Plotting and Scheming with Wavelets," Colm Mulcahy. *Mathematics Magazine*, **69**(5): 323–343(December 1996). (This article received the Carl B. Allendoerfer Award for excellence in exposition.)

"Image Compression Using the Haar Wavelet Transform," Colm Mulcahy. *Spelman Science and Mathematics Journal*, **1**(1): 22–31(1997).

Wavelet Software
and Electronic Resources

Electronic Resources

Wavelet Digest, edited by Wim Sweldens gives up-to-date wavelet information: conferences, publications, software, bibliographies, etc. To subscribe by email, send an empty message to "add@wavelet.org"; to cancel a subscription send a message to "remove@wavelet.org" with your email address as the "subject". The Internet address for the Wavelet Digest is

http://www.wavelet.org/wavelet

Some other web pages of interest:

Fractals: The INRIA (French National Institute for Research in Computer Science and Control) maintains a bilingual fractal page at http://www-syntim.inria.fr.

Fingerprints: A discussion of fingerprint compression can be found at Christopher Brislawn's home page: http://www.c3.lanl.gov/˜brislawn.

JPEG: Links to documents concerning JPEG can be found at www.jpeg.org.

Music: The websites http://www.music.yale.edu/research/index.html and www.fmah.com contain useful information on wavelets and music.

Reproducible Research: A discussion of reproducible research can be found at http://sepwww.stanford.edu.

Steerable Pyramids: Eero Simoncelli has created a web page for the steerable pyramid: http://www.cis.upenn.edu/˜eero/steerpyr.html.

Wavelets: Amara's Wavelet Page http://www.amara.com/current/wavelet.html includes an extensive list of other wavelet web sites, and bibliographies for beginners.

299

Wavelet Software

Following are partial listings of free and commercially available wavelet software. For additional information, try the Wavelet Digest and A. Bruce, D. Donoho, and H.Y. Gao, "Wavelet analysis," *IEEE Spectrum*, Oct. 1996, pp. 26–35.

Anyone publishing research that involves use of software is strongly encouraged to cite the software used, both as a courtesy to the authors and to assist readers who wish to assess research or apply it in their own work. Researchers who have addressed the issue of making computer research easily reproducible include Jonathan Buckheit and David Donoho [Buckheit, Donoho], and Matthias Schwab, Martin Karrenbach, and Jon Claerbout at Stanford University.

"In the mid 1980's, we noticed that a few months after completing a project, the researchers at our laboratory were usually unable to reproduce their own computational work without considerable agony," write the latter [Schwab et al.] ... "Indeed, the problem occurs wherever traditional methods of scientific publication are used to describe computational research. In a traditional article the author merely outlines the relevant computations: the limitations of a paper medium prohibit a complete documentation including experimental data, parameter values, and the author's programs. Consequently, the reader has painfully to re-implement the author's work before verifying and utilizing it." In 1991 they developed guidelines for producing electronic documents that make scientific computations easily reproducible; for more information see http://sepwww.stanford.edu.

Free Software

Note: there is a distinction between software that is in the public domain and software that is available free of charge but is copyrighted, and therefore belongs to the copyright holder, who may set conditions on its use. To the best of the author's knowledge, the following software is available free of charge; for further details, consult the authors or read any copyright information or licensing agreements.

denoise: A 1-D and 2-D wavelet denoising system by Fazal Majid, available by FTP from ftp://math.yale.edu/WWW/pub/wavelets/software/denoise.

EPIC: For wavelet image compression, by Eero Simoncelli and Ted Adelson (M.I.T.), 1989, available at http://www.cis.upenn.edu/˜eero/epic.html. Not state-of-the-art, but the C source code is simple and easy to modify.

EPWIC: For wavelet image compression, by Eero Simoncelli and Rob Buccigrossi, 1996, http://www.cis.upenn.edu/˜butch/EPWIC. State-of-the-art compression based on a simple model for wavelet coefficient statistics.

Fraclab: Christophe Canus, Paulo Goncalvès, Bertrand Guiheneuf and Jacques Lévy Véhel, INRIA (French National Institute for Research in Computer Science and Control). A fractal analysis toolbox including orthogonal and continuous wavelet transforms, wavelet packets and quadratic time-frequency transforms. Available at http://www-rocq.inria.fr/fractales/.

LiftPack: By Gabriel Fernández, Senthil Periaswamy, and Wim Sweldens. Written in C for fast calculation of biorthogonal wavelet transforms using the Lifting Scheme. Both 1-D and 2-D signals of arbitrary size are accepted. All UNIX platforms are supported; with modifications in the installation procedure, the software runs under DOS, Windows 3.1/95/NT, and MAC platforms. A graphical interface in Tcl/Tk/Tix is also provided together with example scripts. Available at http://www.cs.sc.edu/~fernande/liftpack/beta.html. For further information, contact Gabriel Fernández, fernande@cs.sc.edu.

MatlabPyrTools: By Eero Simoncelli, Courant Institute for Mathematical Sciences. Includes Matlab code for building and manipulating Laplacian pyramids, QMF/Wavelets, and steerable pyramids. Data structures are compatible with the Matlab wavelet toolbox, but the code (in C) is faster and includes flexible boundary handling. Available as a gnu-zipped UNIX "tar" file, via anonymous ftp from ftp.cis.upenn.edu in the file pub/eero/matlabPyrTools.tar.gz. Also available at http://www.cis.upenn.edu/~eero.

MatLab Toolbox: For the use of filter bank trees in audio signal processing. Being developed by Frank Kurth and others at the Institut für Informatik, University of Bonn. For availability, check http://theory.cs.uni-bonn.de/~kurth/.

Rice Wavelet Toolbox Release 2.0: A collection of MATLAB mfiles and mex files for 2-band and M-band filter bank/wavelet analysis.These files are available at http://www-dsp.rice.edu/software/RWT/. To get on a mailing list, send an email address to wlet-tools@ece.rice.edu.

SPlus WaveThresh: By Guy Nason, University of Bristol. An add-on package for the statistical package S-PLUS, in two versions: Unix and S-Plus for Windows. Available at http://www.stats.bris.ac.uk/pub/software/wavethresh/WaveThresh.html

WAILI, Wavelets with Integer Lifting: Written in C++ by Geert Uytterhoeven, Filip Van Wulpen, and Maarten Jansen, Catholic University of Leuven in Belgium. Includes use of integer wavelet transforms based on the Lifting Scheme. WAILI is available in source form at http://www.cs.kuleuven.ac.be/~wavelets/ for research purposes only and may not be further distributed. Enquiries about license conditions should be directed to wavelets@cs.kuleuven.ac.be.

WavBox Software: The older Versions 1-3 of WavBox Software, previously distributed free of charge, are no longer available from the author, Carl Taswell. However, WavBox Software Version 1 is still available as Algorithm 735 from the ACM. See C. Taswell and K. McGill, "Algorithm 735: Wavelet Transform Algorithms for Finite- Duration Discrete-Time Signals," *ACM Transactions on Mathematical Software*, 20(3):398-412, September 1994. The software can be downloaded free of charge from http://www.acm.org/calgo/contents/735.gz.

WaveLab: Jonathan Buckheit, Shaobing Chen, David Donoho, and Iain Johnstone, at Stanford University, and Jeffrey Scargle, at NASA-Ames. A library of MATLAB routines for wavelet analysis, wavelet-packet analysis, cosine-packet analysis and

Matching Pursuit. Versions are provided for Macintosh, UNIX and Windows. Extensive documentation. WaveLab has been used in courses at Stanford and at Berkeley; it may be used to reproduce the figures in the authors' published articles, and to redo those figures with variations in the parameters. It was also used in [Mallat 4]. Available at http://playfair.stanford.edu/~wavelab and ftp://playfair.stanford.edu/pub/wavelab.

Wavelet Image Compression Construction Kit: A set of C++ source files; implements a wavelet transform-based image coder for grayscale images. These files are available at http://www.cs.dartmouth.edu/~gdavis/wavelet/wavelet.html.

wplib: Contains the first version of Victor Wickerhauser's own wavelet transform codes, plus utilities that can be used to convert sound and image files between various formats, and early versions of WPLab (see below). It is available by FTP from ftp://math.yale.edu/WWW/pub/wavelets/software/wplib.

XWPL and WPLab: More limited versions of the commercially available Wavelet Packet Laboratory for Windows; they cannot save coefficients. Different versions of XWPL work on a variety of computers but they are all quite limited and all require some Unix flavor plus X-Windows. WPLab runs on two kinds of NeXTs, is less limited and requires Unix and NextStep. XWPL is available by FTP at ftp://math.yale.edu/WWW/pub/wavelets/software/xwpl. WPLab is available at ftp://wuarchive.wustl.edu/~tr-math/software/WPLab3.03.tar.Z.

Commercial Software

Numerical Recipes: The second editions of *Numerical Recipes in C* and *Numerical Recipes in Fortran* include discussions of the discrete wavelet transform, multidimensional transforms, and applications of wavelets to image compression; source code is given in accompanying diskettes. William H. Press, Saul A. Teukolsky, William T. Vetterling, and Brian P. Flannery, *Numerical Recipes in C* and *Numerical Recipes in Fortran*, second editions, Cambridge University Press, 1992. $54.95. Diskettes in C, second edition, for IBM and Macintosh; $39.95. Diskettes in Fortran, second edition, for IBM and Macintosh; $39.95. (All prices as of September 1997) Available from Cambridge University Press, 40 West 20th St., New York, N.Y. 10011-4211, Tel. (800) 872-7423.

S+WAVELETS: Available on Windows and major UNIX platforms, is an extension of the S-PLUS language and graphical data analysis environment. It includes: discrete wavelet transforms (and inverse transforms), multiresolution decomposition and analysis, non-decimated wavelet transforms, time-frequency graphical displays, wavelet packet transforms and local cosine transforms, statistical signal extraction and estimation, Best Basis adaptive choice of transform, Matching Pursuit decompositions, support for 1-D and 2-D data with arbitrary sample sizes, and a full range of boundary correction methods. Data Analysis Products Division, MathSoft, Inc., 1700 Westlake Ave. N., Seattle, WA 98109-9891. Tel. (800) 569-0123, (206) 283-8802; Fax (206) 283-6310; email mktg@statsci.com, http://www.mathsoft.com.

WavBox Software and FirWav Filter Libraries: The original MATLAB wavelet toolbox. The WavBox Software Library Version 4.4 is compatible with MATLAB Professional 4.2 - 5.1, and Student 5; it performs multiresolution analyses of 1-D multichannel signals and 2-D images for arbitrary length and size data. The FirWav Filter Library is compatible with MATLAB Professional 4.2 - 5.1, and Student 4 - 5; it includes the systematized collection of complex and real orthogonal, and biorthogonal Daubechies wavelets. ToolSmiths software products will be provided free of charge to "academic investigators engaged in scientific research and educational activities that promote knowledge, understanding, and public awareness of natural resource conservation." More information and user endorsements are found at http:// www.toolsmiths.com/wavbox.shtml. ToolSmiths, P. O. Box 9925, Stanford, CA 94309-9925, Tel. (650) 323-4336, Fax. (650) 323-5779. Email. info@wavbox.com and info@toolsmiths.com.

Wavelet Explorer: An add-on package to Mathematica; available for all platforms that run Mathematica 2.2 or later, $595. Wolfram Research, Inc. 100 Trade Center Drive, Champaign, IL 61820-7237, Tel. (800) WOLFRAM (965-3726) (U.S. and Canada) or (217) 398-0700; Fax. (217) 398-0747; Email: info@wolfram.com (sales inquiries); http://store.wolfram.com/view/wavelet/.

Wavelet and Filter Banks Design Toolkit: Available for Windows and Mac (as of Aug. 1997, no UNIX). Uses a graphical approach to design filter banks as well as wavelets; users can test their designs for signals on-line. Wavelet and Filter Bank Design, National Instruments Corp. Tel. (800) IEEE488; www.natinst.com.

Wavelet Packet Laboratory for Windows: This package was developed by Ronald Coifman, Kresimir Ukraincik, and Victor Wickerhauser for IBM-PCs, as an interactive tool for finding the "best" representation of a digital signal (using a relatively small number of coefficients). Includes manual, diskette, tutorial, and an applications guide with mathematical background and practical examples; shows how to design specialized algorithms. Requires DOS and Windows 3.1. Available from A K Peters, Ltd., 289 Linden St., Wellesley, MA 02181 U.S.A. Tel. (781) 235-2210; Fax (781) 235-2404; email. service@akpeters.com. List (1998 catalog) $350. Licensing of the computational engine for Wavelet Packet Laboratory for Windows, for any workstation, is handled by FMA&H, 1020 Sherman Ave., Hamden, CT 06514, U.S.A. Tel. (203) 248-8212. Cost as of December 1995, $16,000.

Wavelet Toolbox: This software was written by Michael Misiti, Yves Misiti, Georges Oppenheim and Jean-Michel Poggi, of the University of Paris 11, Orsay. Many exercises in [Strang, Nguyen] use the toolbox. Available from The MathWorks, 24 Prime Park Way Natick, MA 01760-1500. Tel. (508) 647-7000; Fax (508) 647-7001. http://www.mathworks.com/products/wavelettbx.shtml.

WavePak: This software is available for PCs on Windows 3.1/95/NT platforms. Includes "wizard" mode for beginners and a tutorial on wavelets. Time-frequency plots can be zoomed, thresholded, and displayed in color or grayscale; 17 orthonormal wavelets are available, including Beylkin, Coifman, Daubechies and Vaidyanathan wavelets. WavePak is a module of **IDS nVision** and requires it to run. Index Data Systems Ltd., Holtby House, 11 Alexandra Place, Scarborough, North Yorkshire. YO12 5JN, England. Sales: Alan Benetts, 100666.351@compuserve.com Technical: Toby Sharp, idsldev@msn.com.

References

[Adelson] E. Adelson. "Subband Coring for Image Noise Reduction," Internal Report: RCA Sarnoff Laboratories, Princeton, NJ, 1986. Available online at www-bcs.mit.edu/people/adelson, under "publications."

[Adelson, Burt] E. H. Adelson and P. J. Burt. "Image Data Compression with the Laplacian Pyramid," in *Proceedings of the Conference on Pattern Recognition and Information Processing*, Dallas, pp. 218–223, IEEE Computer Society Press, Los Angeles, 1981.

[Adelson et al.] E. H. Adelson, E. Simoncelli, and R. Hingorani. "Orthogonal Pyramid Transforms for Image Coding," *Visual Communications and Image Processing II*, Proc. SPIE, Vol. 845, pp. 50–58, 1987. Reprinted in M. Rabbani, ed., *Selected Papers in Image Coding and Compression*, SPIE Milestone Series, pp. 331–339, SPIE Optical Engineering Press, Bellingham, WA, 1992.

[Alpert] B. Alpert. "A Class of Bases in L^2 for the Sparse Representation of Integral Operators," *SIAM Jour. Math. Anal.*, **24**(1): 246–262. (January 1993).

[Alpert et al.] B. Alpert, G. Beylkin, R. Coifman, and V. Rokhlin. "Wavelet-like Bases for the Fast Solution of Second Kind Integral Equations," *SIAM Jour. Sci. Stat. Comp.*, **14**(1): (January 1993).

[Antoine et al.] J. P. Antoine, D. Barache, R. M. Cesar, Jr., and L. da F. Costa. "Shape Characterization with the Wavelet Transform," *Signal Process.*, **62**(3): 265–290 (1997).

[Antoine et al. 2] J. -P. Antoine, P. Carrette, R. Murenzi, and B. Piette. "Image Analysis with 2D continuous Wavelet Transform," *Signal Process.*, **31**: 241–272 (1993).

[Antonini et al.] M. Antonini, M. Barlaud, P. Mathieu, and I. Daubechies. "Image Coding Using Wavelet Transform," *IEEE Trans. Acoust. Signal Speech Process.*, **1**(2): 205–220 (April 1992).

[Arnéodo et al.] A. Arnéodo, F. Argoul, and G. Grasseau. "Transformation en Ondelettes et Renormalisation," in [Lemarié], pp. 125–191.

[Arnéodo et al. 2] A. Arnéodo, E. Bacry, and J. F. Muzy. "The Thermodynamics of Fractals Revisited with Wavelets," *Physica A*, **213**: 232–275 (1995).

[Arnold, Avez] V. I. Arnold and A. Avez. *Ergodic Problems of Classical Mechanics*, W. A. Benjamin, Inc., New York, 1968.

[Bacry et al.] E. Bacry, A. Arnéodo, J. Muzy, and P.-V. Graves. "Wavelet Analysis of DNA Sequences," in A. F. Laine and M. A. Unser, eds., *Wavelet Applications in Signal and Image Processing III*, Proc. SPIE, pp. 489–498, 1995.

[Barlaud et al.] M. Barlaud, P. Sole, T. Gaidon, M. Antonini, and P. Mathieu. "Pyramidal Lattice Vector Quantization for Multiscale Image Coding," *IEEE Trans. Image Process.*, **3**(4): 367–381 (July 1994).

[Barnsley] M. Barnsley. *Fractals Everywhere*, Academic Press, Inc., Boston, 1988.

[Battle] G. Battle. "Heisenberg Proof of the Balian-Low Theorem," *Lett. Math. Phys.*, **15**: 175–177 (1988).

[Benedetto, Frazier] J. Benedetto and M. Frazier, eds., *Wavelets: Mathematics and Applications*, CRC Press, Boca Raton, FL, 1993.

[Benzi et al.] R. Benzi, S. Gilberto, C. Baudet, and G. Ruiz Chavarria. "On the Scaling of Three Dimensional Homogeneous and Isotropic Turbulence," *Physica D*, **80**: 385–398 (1995).

[Berger, Nichols] J. Berger and C. Nichols. "Brahms at the Piano," *Leonardo Mus. Jour.*, **4**: 23–30 (1994).

[Berger et al.] J. Berger, R. Coifman, and I. Popovic. "Aspects of Pitch-Tracking and Timbre Separation: Feature Detection in Digital Audio Using Adapted Local Trigonometric Bases and Wavelet Packets," in *Proceedings of the 1995 International Computer Music Conference*, Banff, Canada, pp. 280–283.

[Beylkin, Keiser] G. Beylkin and J. M. Keiser. "On the Adaptive Numerical Solution of Nonlinear Partial Differential Equations in Wavelet Bases," *Jour. Comp. Phys.*, **132**: 233-259 (1997) and "An Adaptive Pseudo-Wavelet Approach for Solving Nonlinear Partial Differential Equations," in *Multiscale Wavelet Methods for Partial Differential Equations*, Wavelet Analysis and Applications series, Vol. 6, Academic Press, Boston, to appear.

[Beylkin et al.] G. Beylkin, R. Coifman, and V. Rokhlin. "Fast Wavelet Transforms and Numerical Algorithms," *Comm. Pure Appl. Math.*, **44**: 141–183 (1991).

[Bijaoui] A. Bijaoui. "Wavelets and Astronomical Image Analysis," in [Farge et al.], pp. 195–212.

[Bijaoui et al.] A. Bijaoui, E. Slezak, and G. Mars. "Universe Heterogeneities from a Wavelet Analysis," in [Farge et al.], pp. 213–220.

[Bradley et al.] J. Bradley, C. Brislawn, and T. Hopper. "The FBI Wavelet/Scalar Quantization Standard for Grey-scale Fingerprint Image Compression," *Visual Information Processing II*, Vol. 1961 of *Proc. SPIE*, (Orlando, FL), pp. 293–304, Society of Photo-Optical Instrumentation Engineers, Bellingham, WA, 1993.

[Brislawn] C. M. Brislawn. "Fingerprints Go Digital," *Notices of the AMS*, Vol. 42, No. 11, pp. 1278–1283, Nov., 1995.

[Brislawn et al.] C. M. Brislawn, J. Bradley, R.J. Onyshczak, T. Hopper. "The FBI Compression Standard for Digitized Fingerprint Images," *Proc. SPIE*, Vol. 2847 (Denver, CO), pp. 344–355, Aug. 1996.

[Buckheit, Donoho] J. Buckheit and D. Donoho. "Wavelab and Reproducible Research," in A. Antoniadis and G. Oppenheim, eds., *Wavelets and Statistics*, pp. 53–81, Springer-Verlag, Berlin, 1995.

[Burris et al.] C. S. Burrus, R. A. Gopinath, and H. Guo. *Introduction to Wavelets and Wavelet Transforms: A Primer*, Prentice Hall Engineering, Science & Math., Upper Saddle River, NJ, 1998.

[Burt, Adelson] P. Burt and E. Adelson. "The Laplacian Pyramid as a Compact Image Code," *IEEE Trans. Comm.*, **31**: 482–540 (April 1983).

[Calderón] A. P. Calderón. "Intermediate Spaces and Interpolation, The Complex Method," *Stud. Math.*, **24**: 113–190 (1964).

[Carslaw] H. S. Carslaw. *Introduction to the Theory of Fourier's Series and Integrals*, Macmillan and Co., Ltd., London, 1930.

[Cayley] A. Cayley. "A Memoir of the Theory of Matrices," (first published in 1858), in *The Collected Mathematical Papers*, Vol. 2, pp. 475–496, The University Press, Cambridge, 1889; Johnson Reprint Corporation, New York, 1963.

[Cesar, Costa] R. M. Cesar, Jr. and L. da F. Costa. "Application and Assessment of Multiscale Bending Energy for Morphometric Characterization of Neural Cells," *Rev. Sci. Instruments*, **68**(5): 2177–2186 (May 1997).

[Cesar et al.] R. M. Cesar, Jr., R. C. Coelho, and L. da F. Costa. "Automatic Classification of Retinal Ganglion Cells," in Proc. II Workshop on Cybernetic Vision, IFSC-USP, Sao Carlos, Brazil, IEEE CS Press, pp. 51–56, 1997.

[Chen, Donoho] S. Chen and D. Donoho. "Examples of Basis Pursuit," in A. F. Laine and M. A. Unser, eds., *Wavelet Applications in Signal and Image Processing III*, Proc. SPIE, pp. 564–574, 1995.

[Coifman, Wickerhauser 1] R. R. Coifman and M. V. Wickerhauser. "Wavelets and Adapted Waveform Analysis," in [Benedetto, Frazier], pp. 399–423.

[Coifman, Wickerhauser 2] R. R. Coifman and M. V. Wickerhauser. "Entropy Based Algorithms for Best Basis Selection," *IEEE Trans. Info. Theory.*, **32**: 712–718 (March 1992).

[Coifman et al.] R. R. Coifman, Y. Meyer, S. Quake, and M. V. Wickerhauser. "Signal Processing and Compression with Wavelet Packets, " in Y. Meyer and S. Roques, eds., *Progress in Wavelet Analysis and Applications*, Proc. of the International Conference on Wavelets and Applications, Toulouse, France, Editions Frontières, 1993.

[Couderc et al.] J. P. Couderc, S. Fareh, et al., "Stratification of Time-Frequency Abnormalities in the Signal-Averaged High-Resolution ECG in Postinfarction Patients with and without Venticular Tachycardia and Congential Long QT Syndrome," *Jour. Electrocardiology*, **29**(supplement): 180–188 (1996).

[Cousin] V. Cousin. "Notes Biographiques pour Faire Suite à l'Éloge de M. Fourier," 1831, Archives, Academy of Sciences, Institut de France, Paris.

[Croisier et al.] A. Croisier, D. Esteban, and C. Galand. "Perfect Channel Splitting by Use of Interpolation/Decimation/Tree Decomposition Techniques," in *Int. Conf. on Inform. Sciences and Systems*, Patras, Greece, pp. 443–446.

[Daubechies] I. Daubechies. *Ten Lectures on Wavelets*, SIAM, Philadelphia, 1992.

[Daubechies, Sweldens] I. Daubechies and W. Sweldens. "Factoring Wavelet Transforms into Lifting Steps." Preprint, Bell Laboratories, Lucent Technologies, 1996, to appear in *Jour. Fourier Anal. Appl.* Available online at http://www.cs.sc.edu/~fernande/liftpack/liftbibl.html#factor.

[Davis, Hersh] P. J. Davis and R. Hersh. *The Mathematical Experience*, p. 36, Birkhäuser, Boston, 1980.

[Delambre] J.-B. Delambre. *Rapport Historique sur le Progrès des Sciences Mathématiques Depuis 1789*, Belin, Paris, 1989.

[Delprat et al.] N. Delprat, B. Escudié, P. Guillemain, R. Kronland-Martinet, P. Tchamitchian, and B. Torrésani. "Asymptotic Wavelet and Gabor Analysis: Extraction of Instantaneous Frequencies," *IEEE Trans. Info. Theor.*, **38** (special issue on wavelet and multiresolution analysis): 644–664 (1992).

[DeVore, Lucier] R. A. DeVore and B. J. Lucier. "Fast Wavelet Techniques for Near-Optimal Image Processing," *MILCOM '92*, Vol. 3, pp. 1129–1135, IEEE, New York, 1992.

[Donoho, Johnstone] D. L. Donoho and I. Johnstone. "Ideal Spatial Adaptation via Wavelet Shrinkage," Technical Report No. 400, Stanford University, Department of Statistics, July 1992. (Available in LaTex, DVI, and PostScript by anonymous FTP, from playfair.stanford.edu. Use the command "cd pub/reports", then "ls" for a list of files, and then "get [name of the file]" to obtain a file.)

[Donoho et al.] D. L. Donoho, I. M. Johnstone, G. Kerkyacharian, and D. Picard. "Wavelet Shrinkage: Asymptopia?" (with discussion), *Jour. Royal Stat. Soc.*, Series B, **57**(2): 301–369 (1995).

[Donovan et al. 1] G. C. Donovan, J. S. Geronimo, and D. P. Hardin. "Fractal Functions, Splines, Intertwining Multiresolution Analysis, and Wavelets," in Proc. SPIE, Vol. 2303, pp. 238–256, SPIE, San Diego, July 1994.

[Donovan et al. 2] G. C. Donovan, J. S. Geronimo, and D. P. Hardin. "Intertwining Multiresolution Analyses and the Construction of Piecewise Polynomial Wavelets," *SIAM Jour. Math. Anal.* **27**(4): 1719–1815 (1997).

[Donovan et al. 3] G. C. Donovan, J. S. Geronimo, D. P. Hardin, and P. R. Massopust. "Construction of Orthogonal Wavelets Using Fractal Interpolation Functions," *SIAM Jour. Math. Anal.*, **27**: 1158–1192 (1996).

[Duffin, Schaeffer] R. J. Duffin and A. C. Schaeffer. "A Class of Nonharmonic Fourier Series," *Trans. Amer. Math. Soc.*, **72**: 341–366 (1952).

[Duhamel, Vetterli] P. Duhamel and M. Vetterli. "Fast Fourier Transforms: A Tutorial Review," *Signal Process.*, **19**: 259–299 (1990).

[Einstein] A. Einstein. "Maxwell's Influence on the Development of the Conception of Physical Reality," in *James Clerk Maxwell: A Commemoration Volume 1831-1931*, pp. 66–73, At the University Press, Cambridge, 1931.

[Faraday] M. Faraday. Letter to Maxwell, in *The Life of James Clerk Maxwell*, L. Campbell and W. Garnett, p. 206, Macmillan and Co., London, 1884.

[Farge] M. Farge. "Wavelet Transforms and Their Applications to Turbulence," *Ann. Rev. Fluid Mech.*, **24**: 395–457 (1992).

[Farge et al.] M. Farge, J. C. R. Hunt, and J. C. Vassilicos, eds. *Wavelets, Fractals, and Fourier Transforms*, Clarendon Press, Oxford, 1993.

[Field 1] D. Field. "Relations Between the Statistics of Natural Images and the Response Properties of Cortical Cells," *Jour. Optic. Soc. Amer.*, Series A, **4**: 2379–2394 (December 1987).

[Field 2] D. Field. "Scale Invariance and Self-Similar 'Wavelet' Transforms: An Analysis of Natural Scenes and Mammalian Visual Systems," in [Farge et al.], pp. 151–193.

[Field 3] D. Field. "What is the Goal of Sensory Coding?," *Neural Comp.*, **6**(4): 559–601 (1994).

[Fourier 1] J. Fourier. *The Analytical Theory of Heat*, translated by Alexander Freeman, Cambridge: At the University Press, London, 1878.

[Fourier 2] J. Fourier. Letter to Villetard, 1795, Fourier File AdS, Archives, Academy of Sciences, Institut de France, Paris.

[Frazier] M. Frazier, *An Introduction to Wavelets Through Linear Algebra*, Springer (Undergraduate Texts in Mathematics Series), expected publication 1998.

[Freeman, Adelson] W. T. Freeman and E. H. Adelson. "The Design and Use of Steerable Filters," *IEEE Trans. Pattern Anal. Mach. Intell.*, **13**(9): 891–906 (September 1991).

[Gabor] D. Gabor. "Theory of Communication," *Jour. Inst. Elect. Eng.*, **93**(3): 429–457 (1946).

[Gauss] C. F. Gauss. "Theoria Interpolationis Methodo Nova Tractata," in *Werke*, Vol. 3, pp. 265–330, Königliche Gesellschaft der Wissenschaften, Göttingen, Georg Olms Verlag, Hildesheim-New York, 1866.

[Goldstine] H. H. Goldstine. *A History of Numerical Analysis from the 16th through the 19th Century*, pp. 249–258, Springer-Verlag, New York, 1977.

[Gonnet, Torrésani] C. Gonnet and B. Torrésani. "Local Frequency Analysis with the Two-Dimensional Wavelet Transform," *Signal Process.* **37**: 389–404 (1994).

[Gortler et al.] S. J. Gortler, P. Schröder, M.F. Cohen, and P. Hanrahan. "Wavelet Radiosity," in *Computer Graphics Proc.*, Annual Conference Series, 1993.

[Granlund] G. Granlund. "In Search of a General Picture Processing Operator," *Comp. Graphics Image Process.*, **8**: 155–173 (1978).

[Gratten-Guinness, Ravetz] I. Gratten-Guinness and J. R. Ravetz. *Joseph Fourier, 1768–1830*, MIT Press, Cambridge, MA, 1972. (For another biography of Fourier, see [Herivel].)

[Haar] A. Haar. "Zur Theorie der Orthogonalen Funktionensysteme," *Math. Annal.*, **69**: 331–371 (1910).

[Healy, Weaver] D. Healy, Jr. and J. B. Weaver. "Two Applications of Wavelet Transforms in Magnetic Resonance Imaging," *IEEE Trans. Info. Theory*, **38**(2): 840–869 (March 1992).

[Heideman et al.] M. T. Heideman, D. H. Johnson, and S. C. Burrus. "Gauss and the History of the Fast Fourier Transform," *IEEE Acoust. Speech Signal Process. Mag.*, **1**(4): 14–21 (October 1984).

[Herivel] J. Herivel. *Joseph Fourier: The Man and the Physicist*, Clarendon Press, Oxford, 1975. (For another biography of Fourier, see [Gratten-Guinness, Ravetz].)

[Hermite] C. Hermite. *Correspondance d'Hermite et de Stieltjes*, Vol. 2, Gauthier-Villars, Paris, 1905.

[Holschneider et al.] M. Holschneider, R. Kronland-Martinet, J. Morlet, and P. Tchamitchian. *Wavelets, Time-Frequency Methods and Phase Space*, pp. 289–297, Springer-Verlag, Berlin, 1989.

[Hopper, Preston] T. Hopper and F. Preston. "Compression of Grey-Scale Fingerprint Images," in *Proc. IEEE Data Compression Conference*, Snowbird, UT, pp. 309–318, IEEE Computer Society Press, 1992.

[Hubbard] J. Hubbard. *Differential Equations III*, Springer-Verlag, New York, to appear.

[Hubbard, Hubbard] B. B. Hubbard and J. Hubbard. "Loi et Ordre dans l'Univers: Le Théorème KAM," *Pour la Science*, **118**: 74–82 (June 1993).

[Hubbard, McDill] J. Hubbard and J. M. McDill. "The Binomial Sum of a Fourier Series," preprint.

[Hugo] V. Hugo. *Les Misérables*, Vol. 1, Folio Edition, p. 185.

[Jacobi] C. Jacobi. *C. G. J. Jacobi's Gesammelte Werke*, Vol. 1, Verlag von D. Reimer, Berlin, 1881.

[Karmarkar] N. K. Karmarkar. "A New Polynomial-Time Algorithm for Linear Programming," *Combinatorica*, **4**(4): 373–395 (1984).

[Körner] T. W. Körner. *Fourier Analysis*, Cambridge University Press, Cambridge, UK, 1988.

[Lagrange] L. de Lagrange. *Oeuvres de Lagrange*, Gauthier-Villars, Paris, 1867–1892. (The letter to d'Alembert is in Vol. 13, p. 368.)

[Laplace] P. S. Laplace. *Oeuvres Complètes de Laplace*, Vol. 7, Gauthier-Villars, Paris, 1886.

[Lemarié] P.-G. Lemarié, ed., *Les Ondelettes en 1989*, Lecture Notes in Mathematics, Vol. 1438, Springer-Verlag, New York, 1990.

[Lemarié, Meyer] P.-G. Lemarié and Y. Meyer. "Ondelettes et Bases Hilbertiennes," *Revista Matematica Iberoamericana*, **2**: 1–18 (1986).

[Li, Vitanyi] M. Li and P. M. B. Vitanyi. "Applications of Kolmogorov Complexity in the Theory of Computation," in Alan Selman, ed., *Complexity Theory Retrospective*, pp. 147–203, Springer-Verlag, New York, 1990.

[Lippert et al.] L. Lippert, M. H. Gross, and C. Kurmann. "Compression Domain Volume Rendering for Distributed Environments," *Comp.Graphics Forum*, **16**(3): 95–107 (1997). Available online at ftp.inf.ethz.ch/pub/publications/tech-reports/2xx/263.ps.gz and ftp.inf.ethz.ch/pub/publications/tech-reports/2xx/263.pdf.

[LoPresto et al.] S. M. LoPresto, K. Ramchandran, and M. Orchard. "Image Coding based on Mixture Modeling of Wavelet Coefficients and a Fast Estimation-Quantization Framework," in *Data Compression Conference*, Snowbird, UT, March, 1997, pp. 221-230.

[Lucier et al.] B. J. Lucier et al. "Wavelet Compression and Segmentation of Digital Mammograms," *Jour. Digit. Imag.*, **7**(1):27–38 (1994).

[Mallat 1] S. Mallat. "A Compact Multiresolution Representation: The Wavelet Model," in *Proceedings of the IEEE Computer Society Workshop on Computer Vision*, pp. 2–7, IEEE Computer Society Press, Washington, D.C., 1987.

[Mallat 2] S. Mallat. "Multiresolution Approximation and Wavelets," *Trans. Amer. Math. Soc.*, **315**: 69–88 (1989).

[Mallat 3] S. Mallat. "A Theory for Multiresolution Signal Decomposition: The Wavelet Representation," *IEEE Trans. Pattern Anal. Mach. Intell.*, **11**(7): 674–693 (July 1989).

[Mallat 4] S. Mallat. *A Wavelet Tour of Signal Processing*, Academic Press, Boston, 1998.

[Mallat, Hwang] S. Mallat and W. L. Hwang. "Singularity Detection and Processing with Wavelets," *IEEE Trans. Info. Theory*, **38**(2): 617–643 (March 1992).

[Mallat, Zhong] S. Mallat and S. Zhong. "Characterization of Signals from Multiscale Edges," *IEEE Trans. Pattern Anal. Mach. Intell.*, **14**(7): 710–732 (July 1992).

[Mallat, Zhang] S. Mallat and Z. Zhang. "Matching Pursuits with Time-Frequency Dictionaries," *IEEE Trans. Signal Process.*, **41**(12): 3397–3415 (December 1993).

[Marr] D. Marr. *Vision*, W. H. Freeman, New York, 1982.

[Maxwell] J. C. Maxwell. "Harmonic analysis," in *Encyclopedia Britannica*, Vol. 11, p. 106, William Benton, Chicago, 1968.

[Meyer 1] Y. Meyer. "Principe d'Incertitude, Bases Hilbertiennes et Algèbres d'Opérateurs," in *Séminaire Bourbaki*, No. 145–146, pp. 209–223, Astérisque, Paris, 1987.

[Meyer 2] Y. Meyer. *Wavelets and Operators*, English translation by D. H. Salinger, Cambridge University Press, Cambridge, UK, 1992.

[Meyer 3] Y. Meyer. *Wavelets Algorithms and Applications*, English translation by R. Ryan, SIAM, Philadelphia, 1993.

[Meyer 4] Y. Meyer, ed., *Wavelets and Applications: Proceedings of the International Conference, Marseille, France*, Masson, Paris and Springer-Verlag, Berlin, 1992.

[Meyer 5] Y. Meyer. *Les Ondelettes*, unpublished manuscript.

[Meyer, Coifman] F. G. Meyer and R. R. Coifman. "Brushlets: a tool for directional image analysis and image compression," *Applied and Computational Harmonic Analysis*, **4**: 147–187, Academic Press, Boston, 1997. Available online at http://noodle.med.yale.edu/~meyer/profile.html.

[Mintzer] F. Mintzer. "Filters for Distortion-Free Two-Band Multirate Filter Banks," *IEEE Trans. Acoust. Signal Speech Process.*, **33**(3): 626–630 (June 1985).

[Moser] J. Moser. *Stable and Random Motions in Dynamical Systems*, Princeton University Press, Princeton, NJ, 1973.

[Murenzi] R. Murenzi. "Wavelet Transforms Associated to the n-Dimensional Euclidean Group with Dilations : Signals in More than One Dimension," in J-M. Combes, A. Grossmann, and Ph. Tchamitchian, eds., *Wavelets, Time-Frequency Methods and Phase Space,* Proc. Marseille 1987, pp. 239–246, Springer-Verlag, Berlin, 1989; Second edition, 1990.

[Parisi, Frisch] G. Parisi and U. Frisch. "On the Singularity Structure of Fully Developed Turbulence," in M. Ghil, R. Benzi, and G. Parisi, eds., *Turbulence and Predictability in Geophysical Fluid Dynamics*, Proc. of the International School of Physics, E. Fermi, 1983, pp. 84–87, North Holland, Amsterdam, 1985.

[Pierce, Noll] J. R. Pierce and A. M. Noll. *Signals: The Science of Telecommunications*, Scientific American Library, New York, 1990.

[Poincaré 1] H. Poincaré. "La Logique et l'Intuition dans la Science Mathématique et dans l'Enseignement," in *Oeuvres de Henri Poincaré*, Vol. 11, pp. 130–131, Gauthier-Villars, Paris, 1956.

[Poincaré 2] H. Poincaré. *La Théorie Analytique de la Propagation de la Chaleur*, Gauthier-Villars, Paris, 1895.

[Rioul] O. Rioul. *Ondelettes Régulières: Application à la Compression d'Images Fixes*, Doctoral Thesis, France Telecom CNET, City, March 1993.

[Rioul, Vetterli] O. Rioul and M. Vetterli. "Wavelets and Signal Processing," *IEEE Signal Process. Mag.*, **8**(4): 14–37 (October 1991).

[Ruskai] B. Ruskai, ed., *Wavelets and Their Applications*, Jones and Bartlett, Boston, 1992.

[Ruttimann et al.] U. E. Ruttimann, M. Unser, R. R. Rawlings, D. Rio, N. F. Ramsey, V. S. Mattay, D. W. Hommer, J. A. Frank, and D. R. Weinberger. "Statistical

Analysis of Functional MRI in the Wavelet Domain," *IEEE Trans. Med.Imaging*, to appear.

[Schröder, Sweldens] P. Schröder and W. Sweldens. "Spherical Wavelets: Efficiently Representing Functions on the Sphere," in *Comp. Graphics Proc.*, Annual Conference Series, pp. 161–172, 1995.

[Schwab et al.] M. Schwab, M. Karrenbach, and J. Claerbout. "Making Scientific Computations Reproducible." Available online at http://sepwww.stanford.edu/redoc/cip.html.

[Shannon, Weaver] C. E. Shannon and W. Weaver. *The Mathematical Theory of Communication*, The University of Illinois Press, Urbana, IL, 1964.

[Shapiro] J. M. Shapiro. "Embedded Image Coding Using Zerotrees of Wavelet Coefficients," *IEEE Trans. Signal Proc.*, **41**(12): 3445–3462 (December 1993).

[Siegel] C. L. Siegel. "Iterations of Analytic Functions," *Ann. Math.*, **43**: 607–612 (1942).

[Simoncelli, Adelson] E. P. Simoncelli and E. H. Adelson. "Noise Removal via Baysian Wavelet Coring," in *Proc. 3rd International Conference on Image Processing*, Lausanne, Switzerland. Available at online www-bcs.mit.edu/people/adelson, under "publications."

[Simoncelli et al.] E. P. Simoncelli, W. T. Freeman, E. H. Adelson, and D. J. Heeger. "Shiftable Multiscale Transforms," *IEEE Trans. Info. Theory*, **38**(2): 587–607 (March 1992).

[Sinha, Richards] B. Sinha and K. J. Richards. "The Wavelet Transform Applied to Flow around Antarctica," in [Farge et al.], pp. 221–228.

[Sinha, Tewfik] D. Sinha and A. H. Tewfik. "Low Bit Rate Transparent Audio Compression using Adapted Wavelets," *IEEE Trans. Signal Process.*, **41**(12), 3463–3479 (December 1993).

[Smith, Barnwell] M. J. T. Smith and T. P. Barnwell, III. "Exact Reconstruction for Tree-Structured Subband Coders," *IEEE Trans. Acoust., Speech, Signal Process.*, **34**(3): 431–441 (June 1986).

[Strang] G. Strang. "Wavelet Transforms versus Fourier Transforms," *Bull. Amer. Math. Soc.*, **28**(2): 288–305 (April 1993).

[Strang, Nguyen] G. Strang and T. Nguyen. *Wavelets and Filter Banks*, Wellesley-Cambridge Press, Wellesley, MA, 1996.

[Strichartz] R. Strichartz. "How to Make Wavelets," *Amer. Math. Monthly*, **100**(6): 539–556 (June-July 1993). (A more technical version, that also treats multi-dimensional wavelets, can be found in [Benedetto, Frazier], pp. 23–50.)

[Strömberg] J.-O. Strömberg. "A Modified Franklin System and Higher-Order Spline Systems on R^n as Unconditional Bases for Hardy Spaces," in W. Beckner, A. Caldéron, R. Fefferman, and P. Jones, eds., *Conference on Harmonic Analysis in Honor of Antoni Zygmund*, Vol. 2, pp. 475–494, University of Chicago Press, Chicago, IL, 1983.

[Sweldens] W. Sweldens, "The Lifting Scheme: A New Philosophy in Biorthogonal Wavelet Constructions," in A. Laine and M. Unser, eds., *Wavelet Applications in Signal and Image Processing III*, Proc. SPIE 2569, San Diego, CA, pp. 68–79 (1995).

[Tchamitchian, Torrésani] P. Tchamitchian and B. Torrésani. "Ridge and Skeleton Extraction from Wavelet Transform," in [Ruskai], pp. 123–151.

[Thomson] W. Thomson. "Heat," in *Encyclopedia Britannica*, Vol. 11, Ninth Edition, pp. 578, Charles Scribner's Sons, New York, 1880.

[Torrésani] B. Torrésani. *Analyse continue par ondelettes*, Savoirs Actuels, InterÉditions CNRS Éditions, 1995.

[Unser] M. Unser. "Texture Classification and Segmentation Using Wavelet Frames," *IEEE Trans. Image Process.*, **4**(11): 1549–1560 (November 1995).

[Unser, Aldroubi] M. Unser and A. Aldroubi, "A Review of Wavelets in Biomedical Applications," *Proc. IEEE*, **84**(4): 626–638 (April 1996).

[Unser et al. 1] M. Unser, P. Thévenaz, C. Lee, and U. Ruttimann. "Registration and Statistical Analysis of PET Images Using the Wavelet Transform," *IEEE Eng.Med.Bio.*, 603–611 (Sept/Oct. 1995).

[Unser et al. 2] M. Unser, A. Aldroubi, and M. Eden. "On the Asymptotic Convergence of B-Spline Wavelets to Gabor Functions," *IEEE Trans.Inf. Theory*, **38**(2): 864–872 (March 1992).

[van den Berg] J. C. van den Berg, ed., *Wavelets in Physics*, Cambridge University Press, Cambridge, UK, to appear 1998.

[van der Poorten] A. van der Poorten. "A Proof Euler Missed ... Apéry's Proof of the Irrationality of $\zeta(3)$: An Informal Report," *Math. Intell.*, **1**(4): 195–203 (1979).

[Vetterli, Kovacevic] M. Vetterli and J. Kovacevic. *Wavelets and Subband Coding*, Prentice-Hall, Englewood Cliffs, NJ, 1995.

[Weaver, Healy] J. B. Weaver and D. Healy, Jr. "Adaptive Wavelet Encoding in Cardiac Imaging," in *Proceedings of the Society for Magnetic Resonance in Medicine*, p. 3096, Society for Magnetic resonance in Medicine, Berkeley, CA, 1992.

[Weierstrass 1] K. Weierstrass. Letters, *Acta Math.*, **35**(1912).

[Weierstrass 2] K. Weierstrass. "Über das Problem der Störungen in der Astronomie," Lecture given at the Mathematical Seminar of the University, Winter semester, 1880-81, Mittag-Leffler Institute, Djursholm, quoted in [Moser], pp. 6–7.

[Whittaker] J. Whittaker. *Interpolatory Function Theory*, Cambridge Tracts in Math. and Math. Physics, Vol. 33, 1935.

[Wickerhauser 1] M. V. Wickerhauser. "High-Resolution Still Picture Compression," *Dig. Signal Process.*, **2**(4): 204–226 (October 1992). Available online at fttp://wuarchive.wustl.edu/˜tr-math/papers/dsp.ps.z.

[Wickerhauser 2] M. V. Wickerhauser. *Adapted Wavelet Analysis from Theory to Software*, A K Peters, Ltd., Wellesley, MA, 1994.

[Worring, Smeulders] M. Worring and A. W. M. Smeulders. "Digital Curvature Estimation," *Comp. Vision Graphics Image Process.*, **58**: 366–382 (1993).

[Xu et al.] Y. Xu, J. B. Weaver, D. M. Healy, Jr., and J. Lu. "Wavelet Transform Domain Filters: A Spatially Selective Noise Filtration Technique," *IEEE Trans. Image Process.*, **3**: 747–758 (December 1994).

Index

Barbara Burke Hubbard stumbled into a career as science writer when as a newspaper reporter she was hired by M.I.T. to write about research. Later she won a AAAS-Westinghouse Science Journalism Award for articles on acid rain she wrote for the *Ithaca Journal*. Marriage to mathematician John Hubbard of Cornell convinced her there was more to mathematics than she had glimpsed from high school courses; she now works as a freelance writer specializing in mathematical subjects, and as a contributing editor to the *American Scientist*. She lives with her husband and their four children in Ithaca, New York, and occasionally in France. *The World According to Wavelets*, which she wrote originally in French, is her first book.